# Thermoelastic Stress Analysis

# Thermoelastic Stress Analysis

Edited by

**N Harwood** and **W M Cummings**
National Engineering Laboratory

Adam Hilger
Bristol, Philadelphia and New York

*British Library Cataloguing in Publication Data*
Thermoelastic stress analysis.
   1. Stress analysis. Experimental techniques
   I. Harwood, N.  II. Cummings, W. M.
   620.11230287

   ISBN 0-7503-0075-2

*Library of Congress Cataloging-in-Publication Data*
Thermoelastic stress analysis / edited by N. Harwood and W. M.
   Cummings.
      p.    cm.
   Includes bibliographical references and index.
   ISBN 0-7503-0075-2 (hbk.)
   1. Thermoelastic stress analysis.  I. Harwood, N.  II. Cummings,
W. M.
   TA417.5.T44   1991
   620.1'123—dc20             90-46245
                                  CIP

Published under the Adam Hilger imprint by IOP Publishing Ltd
Techno House, Redcliffe Way, Bristol BS1 6NX, England
335 East 45th Street, New York, NY 10017-3483, USA

US Editorial Office: 1411 Walnut Street, Philadelphia, PA 19102

Typeset by MS Filmsetting Limited, Frome, Somerset
Printed in Great Britain by J W Arrowsmith Ltd, Bristol

# Contents

## 8 The Thermoelastic Analysis of Anisotropic Materials

*R T Potter and L J Greaves*

## 9 Post-processing of SPATE Data

*J T Boyle*

# List of Contributors

**M G Beghi**
*Politecnico di Milano*, Via Ponzio 34/3, 20133 Milano, Italy.

**B R Boyce**
*John Deere Inc*, 400 North Vine Street, Horicon, Wisconsin, USA.

**J T Boyle**
*University of Strathclyde*, 75 Montrose Street, Glasgow, United Kingdom.

**W M Cummings**
*National Engineering Laboratory*, East Kilbride, Glasgow, United Kingdom.

**L J Greaves**
*Royal Armaments Research and Development Establishment*, Chobham Lane, Chertsey, Surrey, United Kingdom.

**N Harwood**
*National Engineering Laboratory*, East Kilbride, Glasgow, United Kingdom.

**A K MacKenzie**
*National Engineering Laboratory*, East Kilbride, Glasgow, United Kingdom.

**J McKelvie**
*University of Strathclyde*, 75 Montrose Street, Glasgow, United Kingdom.

**R T Potter**
*Royal Aircraft Establishment*, Farnborough, Hants, United Kingdom.

**P Stanley**
*University of Manchester*, Oxford Road, Manchester, United Kingdom.

**A K Wong**
*Aeronautical Research Laboratories*, PO Box 4331, Melbourne, Australia.

# Preface

Despite the development over recent years of advanced finite element and computer-aided design programs, designers and engineers are still faced with uncertainties when predicting the strength or durability of a component or structure, particularly for complex geometries.

No matter how sophisticated the theoretical analysis software may be, a lack of precision in the representation of the geometry, boundary conditions, the nature or distribution of the loading, or the physical properties of the material inevitably produce a finite uncertainty in the result, and may lead to considerable inaccuracies in the predicted stresses. The assumptions, compromises, and simplifications which almost invariably have to be used in the design process mean that a purely theoretical approach to stress analysis may lead to a lack of confidence in the accuracy of the model, unless validation has been provided by reliable experimental measurements.

The fact that the demands and expectations placed upon theoretical methods have steadily increased and that successful predictions on complex structures require considerable user expertise has led to the growing acceptance that experimental techniques are not merely an alternative to theoretical analysis but that the measurement of experimental data fulfils a necessary and integrated complementary role in enhancing product quality and reliability. Techniques for estimating stress from experimental measurements are thus of fundamental importance to the engineer who has a duty to produce satisfactory designs in terms of strength, safety and reliability.

Thermoelastic stress analysis is a relatively new technique for the experimental estimation of the stresses produced by the application of dynamic loading to a structure. In many cases it is convenient to apply the method to structures that are undergoing fatigue testing, and the rapidity of the technique means that it has great potential, both in a trouble-shooting role and as a design or theoretical model validation tool applicable to a wide range of engineering industries. The rapid developments over the past few years means that the time is now ripe for a book which explains the current state-of-the-art in the procedures and applications of the thermoelastic technique. Seven of the nine chapters in this book are based upon research using a commercial system (SPATE) which is described in detail in Chapters 1 and 3.

For readers unfamiliar with thermoelastic stress analysis, Chapter 1 provides a straightforward introduction to the technique. The method is placed in its historical context and the basic principles of infra-red detectors are described. The thermoelastic technique is also compared with more established approaches to experimental stress analysis. Calibration procedures are discussed and several examples of stress distributions generated from engineering structures are presented to show the effectiveness of the method.

Chapter 2 gives a detailed description of the principles of the deformation mechanisms which occur when solid materials are subjected to applied loads. Techniques and empirical data for the thermal analysis of strain are presented for a variety of materials and load regimes. This chapter is based upon the analysis of micro-mechanical behaviour using a contacting measurement method rather than the infra-red scanning technique used by most of the contributors to the book. The contacting method is described and compared with the infra-red technique.

In thermoelastic stress analysis it is of critical importance to use the most efficient data acquisition and signal processing techniques. These aspects for the latest version of the commercial system are covered in Chapter 3. Special attention is paid to the problem of noise elimination, and explanations of the electronics and computer hardware, software and sampling methods are provided. The evolution of the technology and the approach to data manipulation, post-processing and presentation are described.

An aspect of thermoelastic stress analysis which had been neglected until fairly recently is the effects of the coatings which are applied, particularly to metallic materials, to provide an enhanced and more uniform infra-red response on the surface of the test structure. Chapter 4 presents a mathematical analysis of heat transfer phenomena, in which it is shown how the coating attenuates the measured response by thermal lag and drag effects. The marked variation of these effects with loading frequency is also described.

The fine spatial resolution and non-contacting nature of thermoelastic stress analysis technology mean that the method has considerable potential for experimental studies of the localised and high-gradient elastic-stress distributions around crack tips. Application of the technology to fracture mechanics under a variety of loading modes is covered in Chapter 5 which describes how stress intensity factors were determined from thermoelastic data acquired from several testpieces, including a large welded joint.

It was recognised during the early stages of the evolution of the technology that it would be a great advantage for practical engineering applications if the thermoelastic technique could be extended to the more complex load waveforms commonly encountered in service. Signal analysis procedures and software are described in Chapter 6 which have been used to estimate stress from structures undergoing random excitation. Several examples of such structures are provided, including components which were behaving modally.

It is in Chapter 7 that the basic principles of the thermoelastic effect are presented and discussed in detail. The thermoelastic theory is derived from the first principles of mechanics and thermodynamics. Experimental validation of the theory, which incorporates higher-order terms than the traditional derivation, is provided. The implications of the theory for the measurement of dynamic stress on composites and high-strength alloys with varying mean-stress levels and for the detection of residual stresses are discussed.

The modern approach to engineering design tends to encourage weight saving whenever possible. It is principally for this reason that composite materials, with their structurally efficient fibre alignment and high strength and stiffness to weight ratios, have attracted much attention and development over recent years. The thermoelastic theory has been extended to such anisotropic materials and this is presented in Chapter 8, together with experimental data acquired from several components made from composite materials.

Chapter 9 presents an overview of the usefulness of thermoelastic data from the point of view of a theoretical stress analyst. Some basic principles of stress analysis are presented and techniques which have been proposed for determining individual principal stresses are reviewed. A joint experimental/ analytical approach is encouraged, whereby the thermoelastic data are employed for the validation of a theoretical model which can then be utilised to separate the principal stress vectors over the scanned area.

N Harwood
W M Cummings

# Glossary

| | |
|---|---|
| $A$ | Cross-sectional area |
| | Calibration constant |
| $a$ | Crack or slot length |
| | Spatial frequency |
| $B$ | Microscopic viscosity coefficient for dislocations |
| $b$ | Breadth |
| | Semi-plate width |
| | Burger's vector of dislocations |
| $C$ | Capacitance |
| | Viscous damping constant |
| | Material coefficient |
| $C_{ijkl}$ | Elasticity tensor for composites |
| $C_\mathrm{p}$ | Specific heat at constant pressure |
| $C_\mathrm{v}$ | Specific heat at constant volume |
| $C_\varepsilon$ | Specific heat under constant strain |
| $c$ | Velocity of light |
| | Semi-depth of testpiece |
| | Boundary point constant |
| $D$ | The responsivity of an infra-red detector |
| $d$ | Diameter |
| | Thickness of an inert layer |
| $E$ | Young's modulus of elasticity |
| $e$ | The emissivity of a surface |
| $F$ | Force |
| $F_i$ | Body force per unit mass |
| $f$ | Frequency |
| | Dissipation function |
| $\Delta f$ | Spectral line spacing |
| $f_\mathrm{n}$ | Noise bandwidth |
| $f_\mathrm{r}$ | Resonant (natural) frequency |
| $f_\mathrm{s}$ | Sampling frequency |
| $f_1, f_2$ | Frequencies at the half-power points |
| $G$ | Modulus of rigidity |

| | |
|---|---|
| $G_{xs}$ | The cross-power spectrum between the load input and the drive signal |
| $G_{xx}$ | The auto-power spectrum of the load reference signal |
| $G_{xy}$ | The complex conjugate of the cross-power spectrum between the response and the reference signal |
| $G_{ys}$ | The cross-power spectrum between the response and the drive signal |
| $G_{yx}$ | The cross-power spectrum between the response and the reference signal |
| $G_{yy}$ | The auto-power spectrum of the thermoelastic response signal |
| $g$ | Gravitational acceleration |
| $H$ | Heat transfer coefficient |
| $H^c$ | An 'unbiased' frequency response function estimator |
| $H_v$ | An alternative frequency response function estimator |
| $H_1$ | The traditional frequency response function estimator |
| $H_2$ | The inverse frequency response function estimator |
| $h$ | Height |
| | Planck's constant |
| $I$ | Second moment of area |
| $i$ | Index in summation |
| i | $\sqrt{-1}$ |
| $J$ | Polar moment |
| $K$ | The bulk modulus of the test material |
| | Stiffness |
| | Attenuation parameter |
| $K_e$ | Thermoelastic parameter |
| $K_n$ | Attenuation parameter of $n$th harmonic |
| $K_0$ | Traditional thermoelastic constant |
| $K_I$ | Mode I stress intensity factor |
| $K_{II}$ | Mode II stress intensity factor |
| $k$ | Thermal conductivity |
| $k_I, k_{II}$ | Non-dimensional stress intensity factors |
| $l$ | Length |
| $M$ | Mass |
| | Bending moment |
| | Moisture content per unit mass |
| $m$ | Gradient of linear plot |
| | Material constant |
| | Schmid's orientation factor for dislocation glide |
| $N$ | Number of cycles |
| | Frame size |
| $n$ | The number of ensembled averages |
| | Number of counts in summation |
| $n_x, n_y$ | Normal to planar boundary |

| | |
|---|---|
| $P_d$ | Power dissipated per unit volume |
| $P_e$ | Elastic part of power per unit volume |
| $P_i$ | Power per unit volume stored in the solid |
| $P_m$ | Total power per unit volume |
| $P_p$ | Plastic part of power per unit volume |
| $P_{te}$ | Thermoelastic effective heat source |
| $P_{th}$ | Sum of effective heat sources |
| $P_v$ | Proper heat source |
| $p$ | Pressure |
| $\dot{Q}$ | Rate of heat production per unit volume |
| $Q_d$ | Energy dissipated per unit volume |
| $Q_e$ | Elastic part of mechanical energy per unit volume |
| $Q_i$ | Heat flux rate |
| | Energy per unit volume stored in a solid |
| $Q_m$ | Total mechanical energy per unit volume |
| $Q_p$ | Plastic part of mechanical energy per unit volume |
| $q$ | Heat generated per unit mass due to moisture absorption |
| $R$ | Radius |
| | Temperature correction factor |
| | Amplitude of lock-in amplifier input |
| | Resistance |
| $r, \theta$ | Polar coordinates |
| $S$ | Specific entropy |
| | Infra-red detector output signal |
| | Sampling time |
| $S_{max}$ | Maximum SPATE signal |
| $S_x$ | The Fourier transform of the digitised load reference signal |
| $S_y$ | The Fourier transform of the digitised thermoelastic response signal |
| $s, s_1, s_2$ | Functions for the Poisson equation |
| $T$ | The absolute temperature of the test surface |
| | Periodic time |
| | Total scan time |
| | Filter time constant |
| | Torque |
| $T_0$ | Reference temperature |
| $\delta T$ | The change in temperature due to a stress change |
| $t$ | Time |
| | Thickness |
| $t_c$ | Thickness of coating |
| $t_s$ | Thickness of substrate |
| $t_x, t_y$ | Boundary tractions |
| $U$ | Internal energy |
| $u, v, w$ | Cartesian displacements |

| | |
|---|---|
| $\dot{u}$ | Velocity in $x$-direction |
| $\ddot{u}$ | Acceleration in $x$-direction |
| $V$ | The RMS amplitude of the thermoelastic response voltage |
| | Volume |
| $v_m$ | Mean velocity of mobile dislocations |
| $W$ | Weight, load |
| $W_p$ | Power |
| $w$ | Work done |
| $w_f$ | Hanning window function |
| $X$ | In-phase component of lock-in amplifier output |
| $x(t)$ | Time-domain function |
| $x, y, z$ | Cartesian coordinates |
| $Y$ | Quadrature component of lock-in amplifier output |
| $Y_I, Y_{II}$ | Geometric correction factors |
| $y$ | Distance from neutral axis |
| $y'$ | Ordinate of line scan |
| $\alpha$ | Coefficient of linear expansion |
| $\beta$ | Angle |
| | Combination of material constants |
| $\Gamma$ | Boundary of region |
| $\gamma$ | Shear strain |
| | Grüneisen parameter |
| $\gamma^2$ | Coherence function |
| $\delta_{ij}$ | Kronecker delta |
| $\varepsilon$ | Axial strain |
| $\varepsilon_1$ | Maximum principal strain |
| $\varepsilon_2$ | Minimum principal strain on a free surface |
| $\varepsilon_e$ | Elastic strain |
| $\varepsilon_{ij}$ | Strain tensor |
| $\varepsilon_p$ | Plastic strain |
| $\varepsilon_{\kappa\kappa}$ | First strain invariant |
| $\delta\varepsilon$ | Change in the sum of orthogonal strains |
| $\zeta$ | Damping ratio |
| $\theta$ | Angle of twist |
| | Phase angle |
| | Temperature difference |
| $\lambda$ | Wavelength |
| $\mu, \lambda$ | Lamé constants |
| | Coating lag factors |
| $\nu$ | Poisson's ratio for the test material |
| $\nu_0$ | Critical Poisson's ratio |
| $\rho$ | The density of the test material |
| $\rho_m$ | Mean density of mobile dislocations |
| $\sigma$ | Axial stress |

| | |
|---|---|
| $\sigma_{\kappa\kappa}$ | First stress invariant |
| $\sigma_{ij}$ | Stress tensor |
| $\sigma_{\mathrm{m}}$ | Mean stress |
| | Bulk stress |
| $\sigma_x, \sigma_y$ | Coordinate normal stresses |
| $\sigma_\theta$ | Thermoelastic–plastic limit stress |
| $\sigma_1$ | Maximum principal stress |
| $\sigma_2$ | Minimum principal stress on a free surface |
| $\Delta\sigma$ | Change in remote applied stress |
| $\delta\sigma$ | Change in the sum of orthogonal stresses |
| $\tau$ | Shear stress |
| | Time constant |
| $\tau_{\mathrm{c}}$ | Time constant for exponential process |
| $\Phi$ | Helmholtz free energy function |
| $\phi$ | Skew angle |
| | Normal stress difference |
| $\chi$ | Thermal diffusivity |
| $\psi_{kl}$ | Coefficients of moisture expansion |
| $\Omega$ | Angle of collection |
| | Boundary element region |
| $\omega$ | Angular velocity |
| $\nabla^2$ | Laplace operator |
| $\Delta$ | Incremental operator |
| $\Sigma$ | Summation operator |
| $*$ | Complex conjugate operator |
| $\|$ | Absolute value operator |

# 1

# An Introduction to Thermoelastic Stress Analysis

## N Harwood, W M Cummings and A K MacKenzie

## 1.1 INTRODUCTION

Structures which are in equilibrium under the action of external forces produce stresses within the material as the structure deforms to resist the application of the loading. Although sophisticated numerical methods for theoretical stress analysis have been developed over the past 30 years, it is still of great interest to engineers to be able to make experimental measurements of stress distributions in structures and components. Such experimental data are vital to the process of improving the structural efficiency of designs and locating stress concentrations which may lead to failure of components. There is, however, no direct way of measuring stress. This chapter provides an introduction to a versatile and rapid technique for stress estimation which has come to prominence relatively recently. The basic principles of stress analysis are not explained in this chapter; Chapters 2 and 9 cover certain fundamental aspects of elasticity or readers may refer to a number of texts, such as Timoshenko and Goodier (1982).

Thermoelastic stress analysis refers to the estimation of the state of stress in a structure by the measurement of the thermal response resulting from the application of a load within the elastic range of the material. The phenomenon of gases changing temperature when subjected to changes of pressure is well known. What is much less familiar, however, is that similar principles apply to solid materials, although the temperature changes are much smaller; in mild steel, for example, rapid loading to just below the yield point produces a temperature change of approximately 0.2°C. The relationship between applied stresses and the corresponding temperature changes in solid materials is known as the thermoelastic effect. Since heat transfer causes these temperature changes to be transitory under the application of static loads, the analysis of thermoelastic response is usually performed under a dynamic load condition of a sufficient frequency to maintain an essentially adiabatic state in the material.

The interest in thermoelastic stress analysis has increased greatly over the past decade due to the appearance on the market of an infra-red (IR) scanning system known as SPATE. SPATE is an acronym which stands for Stress Pattern Analysis by measurement of Thermal Emission. The system has sufficient sensitivity that it is capable of estimating full-field stress patterns from dynamically loaded structures. The SPATE equipment actually measures load-correlated temperature changes on the surface of the material. Under adiabatic conditions the change in temperature at a point in a component is proportional to the change in volume at that position. Since the change in volume is in turn proportional to the first stress invariant (the sum of the principal stresses) at that point, the measured temperature changes can be converted to stress values on the surface of the structure. The stress estimates which are produced are scalar values proportional to the hydrostatic component of the stress tensor.

Thermodynamic theory predicts that, for the great majority of structural materials, tension produces a cooling effect, whereas compressive stresses produce an increase in temperature. This phenomenon can be used to determine the polarities of the estimated stress values.

### 1.1.1 Description of the thermoelastic stress analysis instrumentation

Thermoelastic stress analysis is performed at the National Engineering Laboratory (NEL) using the SPATE 8000 system, the first commercially available instrument to make practical use of the thermoelastic effect for experimental stress analysis applications.

The system is a computer-controlled, non-contacting instrument which may be used to estimate full-field stress patterns in the form of coloured contour maps showing the distribution of the sum of the principal stresses over optically accessible surfaces on structures subjected to dynamic loading. Quantitative stress values at given points are indicated by a 16-colour scale which is displayed on a monitor adjacent to the thermoelastic stress pattern. A flat-bed plotter can also be used to display stress profiles between any two points within the scan area. The scan data may be stored digitally on a dual floppy disk system or down-loaded to a mainframe computer for post-processing.

The test structure must be dynamically loaded at a frequency high enough to ensure that the thermodynamic condition in the component material can be considered to be adiabatic, in which case reversibility is maintained between mechanical and thermal forms of energy. The minimum frequency for an essentially adiabatic state depends on the thermal conductivity of the material and the gradients of the stress fields generated by the loading conditions. For most metallic components it would not be advisable to use a forcing frequency of less than 3 Hz unless corrections could be made for attenuation due to thermal conduction. The standard SPATE system requires the dynamic loading to be in the form of a uniform cyclic waveform. Sinusoidal loading is usually preferred, since noise rejection is more efficient when all the energy of interest is concentrated at a single frequency.

The equipment contains a highly sensitive detector and amplifier/analyser which generates a signal in response to the thermoelastic infra-red flux emitted from a spot on the surface of a structure which is undergoing uniform cyclic loading. In addition to the infra-red detector signal, a clean reference signal taken from a function generator, a load cell, or a strain gauge connected to the test structure must be supplied to the SPATE analyser so that noise may be rejected-out of the measured thermoelastic response and to allow the phase relationship between the reference and response signals to be determined. The sensitivity of the instrument is quoted to be 0.001°C, which equates to a stress resolution of 1 MPa in steel and 0.4 MPa in aluminium. Normal ambient temperature variations do not dominate the thermoelastic output since these changes do not occur at the same frequency as the load reference signal and are thus rejected by the signal analyser. The depth of focus of the lens system means that SPATE is not restricted to flat surfaces but may be used on complex three-dimensional geometries.

A germanium lens in conjunction with motorised horizontal and vertical scanning mirrors, contained within the detector housing, focus the thermoelastic flux from a spot on the structure onto the IR detector. A visual channel aligned with the IR beam allows the operator to see the part of the structure which is being examined, and cross-hairs mark the exact position where the thermoelastic response is being measured. A hand-set connected to the detector housing allows the operator to move the scanning mirrors manually, to set scan limits, and to mark the geometrical position of key points in the scan field.

Before a scan can be performed, the operator must select both a graphics resolution, which determines the number of measurement points in the scan, and also a dwell time per point. The longer the dwell time per point, the greater will be the rejection of random noise in the estimated thermoelastic response and thus the more repeatable will be the calculated value for the stress at each point in the scan. The selection of scan parameters is a compromise between noise rejection, resolution, and available testing time. During a scan the point of focus is moved automatically in equal increments by the scanning mirror motors.

## 1.2 HISTORICAL REVIEW

The phenomenon of a material changing temperature when it is stretched was first noted by Gough in 1805 who performed certain simple experiments using strands of india-rubber. Such elastomers do not of course behave like metals, and the first genuine observation of what is now known as the thermoelastic effect was made by Weber in 1830; Weber noted that a sudden change in tension applied to a vibrating wire did not cause the fundamental frequency of the wire to change as suddenly as he expected, but that the change took place in a more gradual fashion. He reasoned that this transitory effect was due to a temporary change in temperature of the wire as the higher stress was applied. However, this empirical work did not provide any great insight into the causes of the effect, and scientific understanding of the theoretical basis of thermoelasticity had to wait until William Thomson's investigations which started in 1851. Thomson had a very illustrious scientific career and later became better known as Lord Kelvin. Among his many achievements he developed a thermodynamic theory for solid isotropic materials subjected to a generalised stress field. Lord Kelvin is the key figure in the history of thermoelasticity; he published many papers and articles on his scientific work, including the section on elasticity in the ninth edition of the *Encyclopaedia Britannica* (1878).

Temperature changes in a wide range of materials as predicted by Lord Kelvin's theory were confirmed experimentally within about 20% by Joule in 1859, and in 1882 Haga claimed to have confirmed the theory in steel to within 3%. In 1915 Compton and Webster devised a very sensitive temperature measurement system based on the measurement of the resistance changes of a stretched piano wire which formed one arm of a Wheatstone bridge. Experimental data acquired using this apparatus allowed the authors to claim that the theory had been verified to within an average discrepancy of approximately 0.1%. Tamman and Warrentrup (1937) performed experiments on copper, nickel and a carbon steel. They showed that the thermoelastic effect and the reversal in the direction of the temperature change as energy was suddenly released by plastic strain could be used to indicate the yield point in certain metals. In the following year Zener (1938) developed a theory which showed that thermoelasticity is an important factor in the damping of vibrations in polycrystalline materials. In 1950 Rocca and Bever reviewed earlier papers on thermoelasticity and performed further experimental work on nickel and iron. They found that the thermoelastic effect diminishes close to the Curie temperature in ferromagnetic materials. Further advances in the understanding of the thermodynamics of thermoelasticity resulted from the work of Biot (1956) who investigated the entropy changes in deformed materials.

The first scientist to use IR radiometry to estimate the amplitudes of dynamic stresses was Belgen in 1967. In the following year he published a particularly substantial and detailed study of the key factors in non-contacting measurement of thermoelastic response. Although Belgen's investigations were performed with a single-point radiometer, rather than a scanning system, it is surprising that his investigations did not lead more immediately to commercial development of IR radiometry for full-field stress estimation. Perhaps the failure to exploit Belgen's work may be explained simply by the fact that his published material did not reach the right audience or that there was a lack of confidence in IR detection technology and sensitivity at that time.

In 1974, the Admiralty Research Establishment approached Sira Ltd concerning the possibility of determining a relationship between stress and the temperature changes that may be produced by an applied load. Sira confirmed feasibility and over the next four years, with funding from the Ministry of Defence (MoD), developed a laboratory prototype called SPATE for applications research (Mountain and Webber 1978). A joint MoD/Sira patent was granted in 1978. Sira then formed an industrial group project with several prominent British companies in order to develop an industrial model which had an extended signal analysis system and incorporated a colour display. This project was partially supported by the Department of Trade and Industry. At demonstrations of the industrial model to potential users much interest was aroused, and therefore Sira decided to set up a subsidiary company called Ometron Ltd to manufacture and market a commercial version of the system, which was named SPATE 8000 (Oliver *et al* 1982). Since 1982 many papers on applications of thermoelastic stress analysis have been written based on data acquired using the SPATE 8000 system (Enke 1988). Research programmes which have enabled the technique to be extended to random loading and residual stress measurement are described in Chapters 6 and 7, respectively.

In 1987 the SPATE 9000 model incorporating a more powerful computer with improved software was produced. Some of this software was developed by John Deere Inc and the University of Wisconsin. In the same year a complementary laser-interferometer system for vibration measurement (VPI 9000) based on an AERE Harwell design was also developed. The SPATE 9000 model is described in detail in Chapter 3.

While SPATE was undergoing development, scientists were also investigating the suitability of thermographic scanning equipment for full-field stress analysis. The relatively low sensitivity of such thermographic systems means that they have tended to be used for the analysis of very high elastic stresses or plasticity investigations, such as in Jordan and Sandor (1978) or Nayroles *et al* (1981).

## 1.3 INFRA-RED DETECTION

Surface temperature measurements may be made using contacting devices such as thermocouples or thermistors, but such point-measurement techniques are inconvenient to use if full-field information on surface temperature distributions is required. For this reason scanning IR radiometer systems have been developed for thermal imaging. An outline description of IR detectors is given below; a more detailed description of the principles involved may be found in Burnay *et al* (1988).

Certain materials have the property that incident radiant energy may change their electrical characteristics. Such materials may be used as transducers to convert radiant energy into an electrical signal. Various physical mechanisms are involved, but in all such materials electrical noise is also produced. It is the signal-to-noise ratio or, more accurately, specific detectivity (see Hudson 1969), which is the most important consideration when assessing the performance of a detector. In order to measure the very small temperature changes which occur in solids due to rapid stress changes, a very high performance detection system must be used.

The bandwidth in the electromagnetic spectrum between 3 and 14 $\mu$m contains the maximum photon and energy emission for bodies at room temperature. Since this bandwidth lies in the IR region of the light spectrum, sensors which operate within this bandwidth are known as IR detectors. Atmospheric absorption by water vapour and carbon dioxide molecules further restricts the useful bandwidth to two discrete windows at 3–5 $\mu$m and 8–13 $\mu$m. IR detection systems operating in the 3–5 $\mu$m window tend to be used mainly for elevated temperature applications and to have poorer performance than 8–13 $\mu$m systems, especially in low-humidity environments. There are two basic types of IR sensor—thermal detectors or photon detectors.

A thermal detector responds to incident radiation by changing its temperature, which in turn produces a change in resistance (e.g. bolometer), optical properties (e.g. liquid crystals), or electrical field (e.g. pyroelectric device) across the detector. Pyroelectric detectors may be used in thermal imaging systems where they have the advantage of not requiring to be cooled, but their relatively poor sensitivities make them unsuitable for thermoelastic stress analysis.

A photon detector is made from a semiconductor material in which carriers may be excited from a valence energy band into a mobile conduction band by interaction with incident photons of sufficient energy. This means that, unlike thermal detectors, photon detectors have a fairly abrupt upper cut-off in the wavelength to which they respond, since long-wavelength photons are unable to supply sufficient activation to enable carriers to cross the energy gap between the bound and mobile states. Carriers are excited in proportion to the intensity of the incident radiation. Photon detectors require cooling to minimise thermal excitation of the carriers. High-quality photon detectors may be expected to have a quantum efficiency between 60% and 70% (Dennis 1986). The quantum efficiency is the ratio between the rate of carrier events in the detector resulting from the irradiation, and the photon incident rate.

Examples of commonly used photon detector materials are indium/antimonide (InSb), lead/tin/telluride (LTT) and cadmium/mercury/telluride (CMT). Staring† arrays using a $512 \times 512$ element matrix of sensors made from silicides of palladium, platinum or iridium in conjunction with VLSI (Very Large Scale Integration) technology have also been developed recently.

The ternary alloys LTT and CMT have very fast response times (1 $\mu$s). CMT detectors require a bias current to be applied and are more difficult to manufacture than detectors made from other materials, but they have become the most popular type for high-sensitivity applications. CMT may be used in the 3–5 $\mu$m wavelength window, in which case it need only be cooled to $-78°$C, which may be conveniently achieved using the Peltier effect. However, for the measurement of thermal emission at around ambient temperatures, greater temperature resolution is achieved by changing the band-gap energy so that the detector operates in the 8–13 $\mu$m window. This is done by reducing the proportion by mole fraction of cadmium in the alloy from 28% to 20%, and cooling to $-196°$C (77 K). The cooling may be achieved either using liquid nitrogen or compressed air passed through a Joule–Thomson cell. The original SPATE prototype used a LTT detector, but the later SPATE 8000 and most recent 9000 systems use a CMT detector encapsulated in a Dewar cooled by liquid nitrogen. The SPATE system scans much more slowly than a conventional thermal imager in order to achieve a very high sensitivity.

A parameter which must be taken into account when using a non-contacting measurement system based on IR detection is that of the emissivity of the surface of the target, i.e. the proportion of the radiation emitted compared to an ideal 'black body'. It is very important to achieve a uniformly high emissivity ($>0.9$) from a target surface in order to enhance the measured response and reduce thermal reflections from surrounding objects. Since polished metallic surfaces have low emissivities ($<0.1$), they typically require to be spray painted with a uniformly thin layer of high-emissivity paint before reliable measurements can be made. Non-metallic

---

† Where staring means that the detectors each point at the same spot all the time, i.e. there is no scanning system.

surfaces tend to have much higher emissivities and thus may not need to be coated. A detailed analysis of the possible effects of surface coatings on measured thermoelastic response signals is given in Chapter 4.

## 1.4 FULL-FIELD EXPERIMENTAL STRESS ANALYSIS TECHNIQUES

Full-field stress estimation plays a key part in the design of modern structures. Stress analyses of components are performed to improve cost effectiveness and reliability by optimising the use of material, resulting in lighter weight and a design of greater structural efficiency. Experimental stress analysis of engineering structures and components is used extensively in troubleshooting, design assessment, and in the validation of theoretically predicted data.

Strain gauges are the standard method for the estimation of local strains. However, a full-field stress analysis is often required in order to locate both stress concentrations and also areas which may be understressed and thus wasting material. The finite element method is a standard technique for purely theoretical analysis, but for many structures model idealisations are unsatisfactory or too time-consuming to develop. Therefore, an experimental determination of the stress distribution in actual structures under realistic conditions makes a vital contribution to the component design evaluation.

There is an array of full-field experimental techniques available to the designer or engineer investigating service failures, the main ones being outlined below.

### 1.4.1 Photoelasticity

The transmission photoelasticity technique requires the manufacture of a scale model in transparent plastic which then has a beam of plane-polarised light passed through it. Under a static load an interference fringe pattern (i.e. isochromatics) will be visible in the model when viewed in a polariscope. The fringes must be counted from a datum to calculate the in-plane principal stress difference at a point on the surface of the model. This stress may then be related to the corresponding stress in the actual component by use of a simple formula. By means of rotating the polariscope, taking further measurements (i.e. isoclinic evaluation), and going through a very laborious procedure, the individual principal stress vectors may be determined over the surface of the model.

A stress-freezing technique in which the loaded model must be heated until the secondary bonds break down can be used to retain the stress pattern within the model. Slowly cooling the model and then sectioning it may allow a three-dimensional stress analysis to be undertaken.

The resulting fringe patterns need skilled interpretation, and the exercise is of very limited use unless the model is a fully accurate representation of the actual component. Many practical situations such as anisotropy or inhomogeneity are difficult or impossible to model accurately. Moreover, a precise knowledge of the loading conditions is necessary, and this is not always easy to evaluate or simulate.

Problems connected with modelling and load simulation may be avoided by using the reflection photoelastic technique, in which a birefringent plastic coating is bonded to the structure with a reflective adhesive. The polarised light must therefore pass through the coating twice. The procedure for producing a coating is rather laborious and time consuming, particularly for large components, and the technique is subject to errors due to variations in strain through the thickness of the coating. The reinforcement effect of the coating makes this technique unsuitable for thin or low-modulus components unless appropriate correction factors are applied.

Photoelasticity is primarily used for stress analysis under static loads, although dynamic analysis may be undertaken using high-speed photographic techniques.

### 1.4.2 Moiré interferometry

If a linear diffraction grating is bonded to the surface of a component, a laser interferometer containing a reference grid will reveal interference fringe patterns when the component is stressed. The fringes are contours of constant displacement in a particular direction; orthogonal diffraction gratings allow horizontal and vertical fringe maps to be produced. Recent technological developments have permitted sub-micron grating densities to be attained, thus allowing much finer resolution to be achieved compared to traditional moiré techniques. This fine resolution can, however, only be attained over very small areas, which means that moiré is more suited to very detailed stress analysis on limited areas of interest, rather than application to full-scale engineering structures.

The fact that the fringes are in the form of displacement contours rather than stress is a disadvantage for stress analysis. However, the fringes may be digitised and differentiated in a computer in order to produce strain contours. Alternatively, if there is good contrast between the fringes and the background, moiré of moiré may be performed to convert the displacement contours to strain.

The moiré technique is most suited to flat surfaces and has been used, for example, to investigate time-dependent plasticity, to track damage growth in composite materials, to evaluate fracture mechanics parameters, and to indicate anisotropy and the non-uniform nature of residual stress fields. Moiré is unsuited to very severe environments or high-frequency applications.

### 1.4.3 Laser holography

This technique consists of recording photographically the diffraction pattern produced by high-intensity laser illumination of a component, and later reconstructing its image from the diffraction pattern. A holographic image shows effects of perspective and depth when viewed from various angles. Nevertheless, it is not a complete three-dimensional description since it only reveals the parts of the component which were illuminated by the laser. The basic diffraction pattern is produced by recording the interference pattern between the laser beam and an object beam back-scattered from the component on a high-resolution photosensitive plate. Illumination of the photographic plate by the reference beam (i.e. the laser) will decode the diffraction pattern and produce the holographic image. The hologram stores information as to the intensity of the light back-scattered from the component and its phase relative to the laser beam. The phase information is related to the position of the component, and it is this which provides the three-dimensional characteristic of the holographic image. Reconstructing the image with the object in its original position produces an exact overlay of the image on the object. However, if the object has moved slightly or has been deformed, interference fringes may be observed, the number and spacing of which are dependent on the amount of displacement. A pulsed laser may be used for dynamic applications.

Like the moiré technique, the holographic pattern is in the form of displacement rather than stress contours, but it is insensitive to displacements which are other than out of plane. The technique requires complex and expensive equipment which must be set up with great care and completely isolated from extraneous inputs. Under ideal conditions the theoretical accuracy ought to be better than 1 $\mu$m (i.e. similar to the wavelength of the laser). However, practical constraints mean that this exceptional accuracy is unlikely to be achieved.

### 1.4.4 Brittle lacquer

This technique involves the spraying of a varnish-like material on to the surface of a component. Brittle lacquer can be applied to virtually any structure or material. Any cracks in the coating which appear when the component is loaded may be analysed for the direction and magnitude of tensile surface strains. The coating will crack perpendicular to contours of maximum principal strain if the threshold value of the lacquer is exceeded. The standard threshold strain is about 500 $\mu\varepsilon$ but is fairly heavily dependent on temperature and humidity. Therefore, it is advisable to spray calibration bars at the same time and place as the test specimen. Plasticisers are added during formulation in order to produce coatings with a range of threshold strains. Chilling of the coatings may enable the threshold strains to be reduced to 300 $\mu\varepsilon$ or less. Provided that the threshold strain is exceeded, the technique may give an adequate overall picture of the strain distribution and enable stress concentrations to be located.

This is a cheap and useful on-site technique if maximum tensile stresses are of most interest, but it is inherently imprecise, and is primarily used prior to strain gauging in order to optimise strain gauge usage. Brittle lacquer is a one-off technique, i.e. it is unsuitable for situations where repeat measurements are required.

Following fears raised in the United States about the possible carcinogenic properties of certain constitutents of brittle lacquers, a popular brand was withdrawn from the market. The toxic nature of these lacquers means that they should only be used in very well ventilated conditions, and the technician should wear a face mask.

Ceramic coatings are also available for materials which are capable of being heated to about 500°C to fire the glaze. These have the advantages of environmental stability and suitability for operation at elevated temperatures.

### 1.4.5 Stress pattern analysis by thermal emission

Since the SPATE 8000 and 9000 systems are described in section 1.1.2 and Chapter 3, respectively, only a summary of the advantages and disadvantages of SPATE relative to other techniques is presented in this section; the advantages are listed below:

(i) A qualitative full-field stress pattern can be obtained rapidly, particularly if low spatial resolution is required.

(ii) Only minimal surface preparation is required (i.e. cleaning and spraying with high-emissivity paint). On many non-metallic materials painting may not be necessary.

(iii) Complex geometries or features (e.g. welds) can be analysed almost as easily as flat surfaces, since the IR emission is fairly independent of viewing angles of up to 60° from normal (Webber 1987), and the instrument has a wide depth of focus.

(iv) A spatial resolution of 0.5 mm can be achieved over a scan area of approximately 100 mm × 100 mm. The minimum resolution inevitably becomes coarser as the scan area is increased. A range of display resolutions is available for selection by the operator before the commencement of scanning.

(v) The stress sensitivity of the instrument is better (1 MPa in steel) than can usually be achieved using other full-field techniques.

(vi) The technique is non-contacting.

(vii) The field of view is altered simply by moving the detector assembly and/or the limits of the scanning mirrors.

(viii) The data is in the form of stress rather than displacement contours.

(ix) An easy-to-interpret colour-coded pattern is displayed on a VDU; there is no need to count fringes. Graphs of stress profiles either in real time or from stored area scan data can also be produced on a dedicated plotter.

(x) The stress patterns may be archived on a dedicated floppy disk unit or transferred to a mainframe for detailed examination, post-processing or comparison with finite element data. There is great flexibility in the storage, reduction, analysis and display of data.

(xi) The technique is applied to actual full-scale structures, no modelling being required, and may be used at elevated temperatures.

(xii) The standard system is capable of extension to variable-amplitude, service loading conditions (see Chapter 6).

The SPATE system does also have certain disadvantages, as outlined below:

(i) A fairly high initial capital outlay is required.

(ii) A dynamic load must be applied to the test structure.

(iii) Principal stress vectors are not measured. The output data are in the form of the sum of the principal stresses (i.e. isopachic contours).

(iv) Displacements normal to the viewing angle may produce edge errors and smear the stress estimates. Metallised mirrors may be used to minimise this effect.

(v) Calibration from material properties requires a precise knowledge of physical data which may be difficult to obtain for the component material, and therefore it is advisable either to attach a strain gauge rosette to the component or to manufacture a special calibration testpiece from the component material in order to obtain a reliable quantitative stress pattern.

(vi) Significant ambient temperature changes affect the accuracy of the calibration, due mainly to the strong temperature dependence of radiant energy flux intensity.

(vii) Normally only surface stress can be estimated, although it is possible in principle to use non-adiabatic loading conditions to calculate internal stresses (Lesniak 1988).

(viii) Optical access is required, via mirrors if necessary.

### 1.4.6 Frictional strain gauge and contacting probe

The frictional strain gauge is not strictly a full-field technique, but it is a cheap, fairly rapid and simple method of estimating surface strain distributions on certain structures. The device takes the form of a hand-held probe (figure 1.1) which contains an interchangeable 6 mm strain gauge mounted

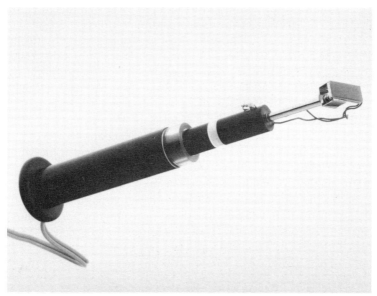

**Figure 1.1** Frictional strain gauge.

on a piece of rubber and held in position by a small clip and spring. The probe and gauge are readily portable and may be pressed against a structure which is undergoing cyclic loading. The conditioned output signal may be read as peak-to-peak or RMS strain on suitable instrumentation. The polarity of the strain estimate is indicated by the phase relationship between the strain signal and a drive or load signal reference.

The probe has been found to be particularly useful when used in conjunction with SPATE to separate principal stress vectors and thus aid in the interpretation of the thermoelastic stress patterns. Uses of the frictional strain gauge have been described by Harwood (1988).

A system based on a roving probe has also recently been developed in the UK. The system is known as VU-strain and uses a probe containing a full-bridge clip gauge extensometer fitted with point knife edges which provide a robust system which is particularly suitable for use on rough surfaces. The probe may be magnetically clamped for use in areas where access is difficult or hazardous. It is more expensive to buy than a frictional strain gauge system, but significantly cheaper than most of the full-field techniques described earlier.

## 1.5 CALIBRATION

There are two standard methods of calibrating thermoelastic data from isotropic, homogeneous materials: a simple component with an exact analytical solution for the stress distribution may be manufactured from an identical material to that of the test structure and then loaded at the same frequency (Stanley and Chan 1985b); or a strain gauge rosette may be bonded to the structure and used to obtain a calibration factor relating the sum of the principal stresses determined from the rosette outputs to the amplitude of the thermoelastic response signal measured adjacent to the strain gauges. Due to the invariant nature of the sum of the principal stresses it is not strictly necessary to use a rosette for the calibration; a T-gauge oriented at an arbitrary angle will provide sufficient data. However, a rosette supplies additional useful information about the stress field at the calibration point for very little extra effort.

A third method of calibration, which may be used when the material properties of the test structure and the IR detection parameters are reliably known, is to substitute values in the theoretical equation for the sum of the principal stresses (see equation (1.4)) which is derived in section 1.5.1.

### 1.5.1 Thermoelasticity theory

The relationship between thermal stress and strain in an isotropic elastic element is given by the following equation (Timoshenko and Goodier 1982)

$$\delta\varepsilon = \frac{(1-2v)\delta\sigma}{E} + 3\alpha\delta T \tag{1.1}$$

where $\delta\varepsilon$ is the sum of orthogonal strain changes, $\delta\sigma$ is the sum of orthogonal stress changes (Pa), $E$ is the Young's modulus of the material (Pa), $v$ is Poisson's ratio, $\alpha$ is the coefficient of thermal expansion ($K^{-1}$), and $\delta T$ is the change in temperature (K). Note that $\delta\varepsilon$ and $\delta\sigma$ are invariants, i.e. they are independent of the co-ordinate axes of measurement. This means that they are scalar rather than vector quantities.

A thermodynamic analysis of the entropy changes in a stressed, elastic element under adiabatic conditions produces the equation (Biot 1956)

$$\delta T = \frac{-3T\alpha K\delta\varepsilon}{\rho C_v} \tag{1.2}$$

where $T$ is the absolute temperature (K), $K$ is the bulk modulus (Pa), $\rho$ is the density ($kg\,m^{-3}$), and $C_v$ is the specific heat at constant volume ($J\,kg^{-1}\,K^{-1}$). Note that $\delta\varepsilon$ is the volume dilation, i.e. if there is no volumetric change in the element (e.g. pure shear), then there will be no change in temperature under elastic conditions.

By making use of the known relationship (Rogers and Mayhew 1980) between $C_v$ and $C_p$, the specific heat at constant pressure, equations (1.1) and

(1.2) may be combined to form the following basic equation describing the thermoelastic effect (Darken and Curry 1953):

$$\delta T = \frac{-\alpha T \delta \sigma}{\rho C_p}. \tag{1.3}$$

Equation (1.3) is only valid if adiabatic conditions prevail, the material is isotropic, and the variation of $\alpha$ and all elastic constants with temperature can be neglected. Any temperature dependence of these constants is normally considered to be a second-order effect in SPATE applications. However, if the temperature dependence of these constants is significant, a varying mean-stress level may lead to a non-linear thermoelastic response, as described in Chapter 7. Although the mean stress effect has been reported by Machin *et al* (1987) as being significant in certain high-strength aluminium and titanium aircraft alloys, experimental investigations at NEL have indicated that for carbon steel components this effect may be ignored for most practical cases. Note that the mean-stress effect may be considerable in composites, as mentioned in Chapter 7.

The estimation of stress using the SPATE system requires that equation (1.3) be modified to include certain other parameters specific to IR detection techniques, resulting in the following working equation

$$\delta \sigma = \frac{-DVR\rho C_p}{\alpha T e} \tag{1.4}$$

where $D$ is the responsivity of the SPATE detector ($K\,V^{-1}$), $V$ is the RMS amplitude of the thermoelastic response voltage at the operating frequency (V), $R$ is a correction factor which compensates for temperature-dependent changes in radiation intensity and wavelength (Rogers and Mayhew 1980), and $e$ is the surface emissivity.

The target-temperature correction factor $R$ is based primarily on the Stefan–Boltzmann law, modified to take account of photon flux over a limited bandwidth (Dennis 1986), and is supplied in graphical form in the operating manual for the SPATE equipment ($R = 1$ at 20°C); the manual also quotes a figure for the responsivity $D$. Equation (1.4) presumes that there is neither significant absorption of the 8–13 $\mu$m IR radiation by the atmosphere nor significant attenuation of the surface thermal response by conduction to and through a paint coating.

Although the small ambient temperature variations which may occur during a scan are normally neglected in most laboratory environments, it should be noted that photon flux is strongly dependent upon the absolute temperature of the target, both in terms of total emittance and wavelength. The wavelength in microns at which maximum photons are emitted is given by dividing 3670 by the absolute temperature (Burnay *et al* 1988). Although many more photons are emitted at elevated temperatures, the shift in wavelength may produce a significant decrease in the signal-to-noise ratio of the photon detector. Moreover, atmospheric absorption may increase significantly, e.g. water vapour will absorb many photons emitted from a target at 300°C. Photons emitted at a given temperature do, however, have a considerable spread of wavelengths and therefore many will be transmitted, even if the maximum emittance is centred on an atmospheric absorption wavelength. Nevertheless, it would be advantageous to use IR equipment for stress estimation around target temperatures at which the wavelength for maximum photon flux lies well within an atmospheric transmission window, e.g. ambient to about 170°C, 460–540°C or 620–950°C. Enke (1989) has discussed the problems that may be encountered in the application of SPATE to elevated temperature testing. He found that the minimum resolvable stress amplitude was almost independent of the temperature of the target.

The thermodynamic details of thermoelasticity are presented in much more detail in Chapter 7. Anisotropic materials are discussed mainly in Chapter 8. The thermoelastic theory only applies to elastic loading conditions. At stress levels sufficient to cause plastic conditions, dynamic loading will produce a hysteresis loop which results in a substantially greater temperature change than is caused by elastic strain (Jordan and Sandor 1978). Nevertheless, such plastic strain often produces 'shakedown', in which the relief of peak stresses resulting from plastic flow allows reversible elastic conditions to be reached within a short time. Thermoplasticity is discussed in detail in Chapter 2.

### 1.5.2 Calibration testpieces

Beam and disc specimens are popular for full-field stress calibration; both these components have a theoretically calculable varying stress field. A SPATE area scan measured on a steel beam subjected to four-point bending is shown in figure 1.2. The parallel, evenly spaced stress contours predicted by simple beam theory may be seen clearly in the stress pattern. The slope of the stress profile across the depth of the beam may be used to obtain a calibration factor to convert thermoelastic data to quantitative stress values for a given material (Stanley and Chan 1985b).

When a disc specimen is subjected to two diametrically opposed point loads, a biaxial stress field is produced. A disc testpiece is simple to load and has a central area of relatively uniform stress (see figure 1.3) remote from any edges which tend to produce spurious data in SPATE scans. Both the beam and disc components have the drawback that they require a compressive static preload, which effectively halves the permissible dynamic load range. The disc has no out-of-phase area in the stress pattern and it has been shown by Timoshenko and Goodier (1982) that there may be a significant departure from the theoretical solution, due to the presence of contact forces at the load points. On the positive side, the disc specimen is particularly suited to investigations of anisotropic materials, since it can readily be rotated to any orientation relative to the load axis.

The practical experience of SPATE users has led many of them towards the use of the axially loaded plate with a central hole for a more qualitative assessment of performance on a given test material. However, the necessity of applying exactly axial loading to conform with the theoretical solution makes the hole-in-plate specimen somewhat less attractive for precise quantitative calibration than the beam or disc.

A calibration component (figure 1.4) has been designed and manufactured at NEL. It consists of a testpiece with a 150 mm long parallel central section, 50.8 mm wide and 12.7 mm thick. Universal joints have been bolted to each end in order to minimise any bending moments which may be applied by the loading train. The use of a component with a nominally uniform stress field makes the calibration procedure very straightforward. Spurious signals produced by noise, surface irregularities or uneven paint application are also more readily apparent.

The testpiece was designed to fit the hydraulic grips in a 250 kN servo-hydraulic test machine. The bearing in the joints consists of a chrome-plated steel ball in a ptfe (polytetrafluoroethylene, commonly known as Teflon) liner. The joint assembly has a peak load rating of 150 kN under dynamic loading conditions, and allows a rotation of the testpiece relative to the joint of at least 6°. The intention of this design was to produce a stress field

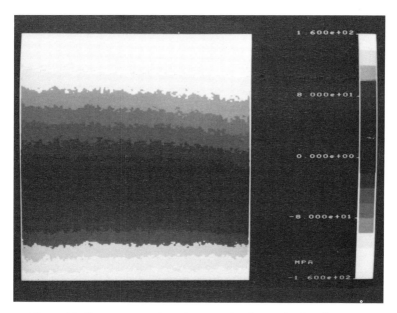

**Figure 1.2** Stress pattern from beam under four-point bending.

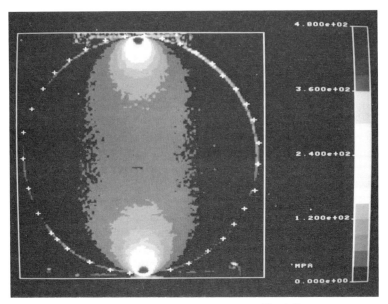

**Figure 1.3** Stress pattern from 'Brazilian' disc.

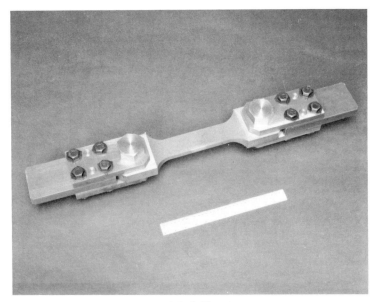

**Figure 1.4** Calibration.

uniform to within 5% over the central section. The average thermoelastic output signal over this area would then be related to the stresses determined from the load cell output, with the axiality being checked using strain gauges.

### 1.5.3 Experimental thermoelastic data and theoretical predictions for a carbon steel

In order to check the axiality of the loading, a 45° strain gauge rosette was bonded to the surface of the centre of one face of the testpiece and a single gauge was bonded to each edge. The gauge installation and the calibration of the load cell and the strain gauge conditioning instrumentation were checked before any measurements were taken. A frequency of 10 Hz was initially selected for the calibration check since this frequency is above the non-adiabatic bandwidth and is low enough to avoid significant signal attenuation due to conduction to and through a thin surface coating (McKelvie 1987).

All the axial gauges agreed to within ±3% and were consistent with the load cell output, assuming a Young's modulus and Poisson's ratio of 211 GPa and 0.29, respectively (Woolman and Mottram 1964). The load cell indicated that the average stress around the central section was 62 MPa peak to peak.

A SPATE area scan (figure 1.5) was performed in order to estimate the stress distribution over the total length of one side of the testpiece. The

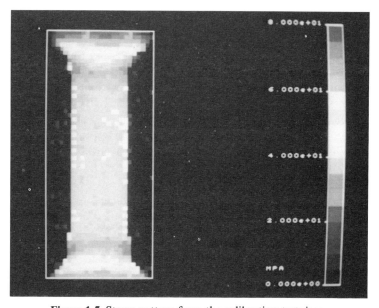

**Figure 1.5** Stress pattern from the calibration testpiece.

unsmoothed scan data show the fairly uniform nature of the stress distribution along the parallel section together with the higher-stressed areas associated with the fillets which blend the central section of the testpiece into the thicker section to which the bearings are attached.

By substitution of known values into the theoretical equation (1.4) relating stress to material properties, the theoretically predicted stress may be compared with that measured by the load cell.

The carbon steel (En 8) used for the calibration investigation contains the following constituents: 0.4% carbon, 0.7% manganese, 0.1% silicon and trace elements of sulphur, phosphorous and nickel. Quoted material properties values (Woolman and Mottram 1964) for En 8 are listed below:

$$\alpha = 11.3 \times 10^{-6} \, \text{K}$$
$$\rho = 7850 \, \text{kg m}^{-3}$$
$$C_p = 480 \, \text{J kg}^{-1} \text{K}^{-1}.$$

The ambient temperature during the calibration was 26°C, i.e. 299 K.

Ometron Ltd quote the responsivity of the detector system in the NEL SPATE head to be 9.0 K V$^{-1}$ and the temperature correction factor to be 0.94 at 26°C. These values were determined using the specially developed radiometric calibration equipment described by Oliver and Webber (1984).

Signal enhancement was achieved by spraying a very thin layer of high-emissivity paint onto the testpiece before any SPATE measurements were taken. The emissivity of the paint used for this investigation (RS matt black) has been measured by Ometron Ltd to be 0.93. Heat transfer theory predicts that there will be no significant attenuation due to coating thermal lag at 10 Hz if a thin layer ($\simeq 20 \, \mu$m) of suitable paint such as RS matt black is used. Paint coating thicknesses on metallic substrates may be measured using an induction gauge.

At the detector stand-off distance of 0.6 m used in the carbon steel calibration there will be no significant absorption of IR radiation by the atmosphere. However, SPATE operation in normal environments requires a film pellicle to be inserted into the detector aperture to prevent the ingress of dust, and this is known to attenuate incoming signals by 14% (Brown 1987). Therefore, if a pellicle is used, a correction factor of 1.16 must be inserted into the theoretical equation (1.4), since the detector responsivity figure quoted previously was determined without the pellicle in position.

The average thermoelastic response voltage measured over the central area was 5.15 mV. This voltage is negative, since tension produces a temperature decrease in steel and in the great majority of structural materials. Therefore, substitution of the above values into equation (1.4) gives

$$\delta\sigma = \frac{-9.0 \times -5.15 \times 10^{-3} \times 0.94 \times 7850 \times 480 \times 1.16}{11.3 \times 10^{-6} \times 299 \times 0.93}$$

$= 60.6$ MPa.

The above value compares with a stress of 62 MPa determined from the load cell, a difference of approximately 2.3%.

The calibration for a carbon steel described in this section is a single-frequency calibration which may not be valid over a wide bandwidth. The SPATE response is known to fall-off at low frequencies. This phenomenon may be due to attenuation produced by non-adiabatic conditions, or relatively poor low-frequency detector response, or the AC-coupling of the detector output. At frequencies higher than 30 Hz consideration must be given to the likelihood of significant effects due to paint coatings.

The low-frequency attenuation was investigated by measuring the amplitude of the thermoelastic response signal from the carbon steel calibration testpiece at discrete frequencies over a bandwidth from 0.5 to 10 Hz. The results are presented in figure 1.6 together with a graph showing the phase lag between the thermoelastic response signal and the load cell output. The RMS voltage of the thermoelastic signal was converted to stress using the calibration factor previously determined at 10 Hz.

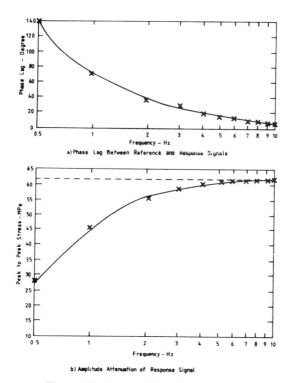

**Figure 1.6** Attenuation at low frequency.

### 1.5.4  A consideration of possible errors in the previous calibration

Although no error bands are given for the calculations presented in the previous section, it is important to consider the effects of possible variations in quoted parameters or the random or systematic errors inherent in experimental techniques.

On the load range used for the acquisition of the experimental data, the NEL test machine and associated instrumentation is certificated as grade 2 according to BS1610, i.e. within 2% mean error in the applied static force. British Standard procedures for dynamic load calibration have not yet been agreed. It was presumed that the static calibration using a full-bridge strain gauge load cell did not vary significantly for dynamic conditions within the low end of the servo-hydraulic range, i.e. inertia errors may be neglected (Dixon 1988). Nevertheless, an important factor to consider when making dynamic measurements is the repeatability of the test control and instrumentation systems. Although great care was taken to minimise system noise and achieve a stable dynamic load reading, a realistic error band on the dynamic load amplitude is unlikely to be much lower than 3%. Estimates of the cross-sectional area of the testpiece using a micrometer give a negligible error compared to other factors, since slip gauges were used to check the accuracy of the instrument.

Combining the above accuracy values in quadrature (Hayward 1977) shows that the measured dynamic stress estimate from the load cell has an error band of $\pm 3.6\%$, assuming that the static load cell calibration is valid for dynamic conditions.

An assessment of the accuracy of the stress predicted from material properties and the SPATE system is more difficult than the load cell based estimate described previously owing to the lack of error band information quoted in reference data.

Single-point measurements on the NEL instrument indicated that the ratio of SPATE response voltage to dynamic load tended to show an instability of typically 3% over a time span of one hour. The 3% error band estimated for the repeatability of the dynamic load will also carry over to the SPATE measurement. The responsivity of the detector is calculated by the manufacturer using a least-squares linear fit of experimental data measured by exposing it to chopped radiation from a calibrated source. This procedure is estimated to give an accuracy within 5% (Webber 1988). The other main parameters where errors could occur are in the quoted values for the specific heat and the coefficient of linear expansion. There tends to be some variability in such material properties quoted for similar materials (Tennant 1971), but the values used in this paper were obtained from the standard reference book for En steels and are very similar to those which are generally accepted for carbon steels. Data presented in ASTM Standard E831 indicate that the accuracy of the coefficient of linear expansion may be expected to be within 5%. No such data is provided in the relevant standard for specific heat

(D2766), but it is reasonable to presume that a figure of 5% would also be appropriate for this parameter. Emissivity may be estimated with an accuracy of 2% (Webber 1988). Estimates of density are not likely to be significantly in error since both mass and volume can be measured very accurately. Any error in the temperature measurement will be insignificant since absolute temperatures are being used in the calculation.

Quadrature combination of the above factors leads to an error band of approximately ±10% on the estimation of the stress predicted from 'known' material properties and SPATE measurements, assuming no significant paint coating thermal lag/drag errors. For many materials, however, the lack of confidence in the accuracy of material properties and the more systematic nature of the errors make this form of calibration relatively unreliable, although it is useful for corroboration of other techniques.

Combination of the relevant factors described above shows that the use of a standard calibration testpiece should give an error band on the SPATE calibration factor within ±6%, which would be acceptable for most engineering applications.

In many cases it is not possible or it is too time-consuming to manufacture a calibration testpiece, and therefore a strain gauge rosette is commonly used for calibration. Under laboratory conditions Chalmers (1980) has shown that strain gauges may be expected to give an accuracy within 3%. Therefore, the use of strain gauges introduces an extra possible error which makes this approach slightly less reliable than a calibration testpiece. Quadrature combination of the relevant factors gives an error band within 7%.

The above figures assume that the surface temperature of the structure does not change significantly during the stress estimation and calibration procedure. Calibration graphs supplied with the SPATE 8000 instrumentation indicate that a 2°C change close to room temperature will introduce an error in the calibration factor of approximately 2%, although it is of course possible, in principle, to compensate for this effect.

A qualitative assessment of the reliability of SPATE data may be made by referring to the unsmoothed full-field stress pattern shown in figure 1.5. The parallel section of the testpiece, where the stress is fairly uniform, consists predominantly of a single colour with a range of ±1 colour (1 colour being approximately 6% of full scale). Most of the pixels which are in error lie near the edge of the testpiece where it is more difficult to make an accurate estimate of the thermoelastic response. Readers will find a more general discussion of sources of error, particularly from a signal processing point of view, in Chapter 3.

## 1.6 FAILURE CRITERIA

A criticism which is commonly levelled against SPATE data is that the stress patterns are in the form of the sum of the principal stresses and that such data are not useful to stress analysts. The sum of orthogonal stresses is known as the first stress invariant and it is proportional to the hydrostatic stress. Since it has been shown experimentally (Drucker 1967) that materials without voids are able to withstand enormous hydrostatic pressures, it is clear that data in this form cannot be used to predict the failure of a structure. In fact, in stress analysis for failure prediction under multi-axial conditions it is commonplace to subtract the hydrostatic component from each stress tensor in order to determine the so-called deviatoric stresses which are important in producing yielding.

For very brittle materials, which are weak in tension due to the presence of flaws, the maximum principal stress is usually considered to be the most suitable yield criterion. However, for ductile materials, the Tresca (maximum shear stress) and the von Mises (shear strain energy) yield criteria have been shown to be more accurate, since they reflect the planar shearing failure mode common to ductile materials and, indeed, to brittle materials in compression. The von Mises criterion is usually preferred since it is a continuous function and all three principal stresses are included in the determination of the effective stress, whereas the Tresca criterion presupposes that the intermediate principal stress makes no significant contribution to failure, an assumption which may not be generally valid. Formulae for failure criteria and a detailed discussion of their use may be found in Timoshenko (1956).

As mentioned in section 1.4.1, the photoelastic technique produces data in the form of the principal stress difference, i.e. the diameter of the Mohr's circle rather than the centre which is determined by SPATE. Since the principal stress difference is proportional to the maximum shear stress, at first sight this may appear to equate to the Tresca failure criterion. However, it is the maximum shear stress parallel to the surface which is being estimated, and this will only be equivalent to the Tresca failure criterion if the principal stresses in the plane of a free surface are of opposite polarity. If the principal stresses have the same polarity on a free surface, the maximum shear stress acts normal to the surface, in which case the Tresca criterion becomes equivalent to the maximum principal stress. Therefore, photoelastic fringe patterns may be misleading for predicting failure in biaxial stress fields, unless the laborious procedure for separating the principal stress vectors is performed.

Like the sum of the principal stresses, the von Mises effective stress is an invariant of the stress tensor. However, it represents the distortion of an element rather than the volumetric change associated with the first invariant. The invariance property is not the only link between the von Mises stress and the sum of the principal stresses; the normal and shear stresses on an octahedral plane (see Fung 1965) are proportional to the principal stress sum and the von Mises stress respectively. Although linked in this way, these two stress scalars are completely independent.

It should be noted that the principal stress difference and the von Mises effective stress are always positive, i.e. data presented in this form do not give any information about whether the biaxial stress field is predominately tensile or compressive. This is in contrast to SPATE data which do contain stress polarity information, which in many cases can be a considerable advantage.

Although failure criteria do produce different effective stress values from the sum of the principal stresses in a biaxial stress field, the highest stresses in components are commonly at a free edge where the stress field is uniaxial. Practical experience using the SPATE system at NEL has indicated that many structures have a predominant principal stress in the main areas of interest, and thus thermoelastic data are more useful for structural assessment than might at first be supposed. The usefulness of thermoelastic data for stress analysis and techniques for principal stress separation are discussed in Chapter 9.

## 1.7 APPLICATIONS OF THE SPATE 8000 SYSTEM AT NEL

Applications of the SPATE system at NEL to date cover most of the major industries, including transport, civil engineering, offshore engineering, medical, and power generation. Several of the structures were examined as part of our commercial services, and therefore the information that can be released is limited.

As an example of a structure taken from the NEL test programme which illustrates some of the important capabilities of thermoelastic stress analysis, figure 1.7 shows a large, steel marine chain link. This link was loaded sinusoidally in tension in a servo-hydraulic test machine. Data from this standard link were compared with that from a new design, with the intention of optimising the design to improve the fatigue properties in service. The stress pattern measured over the whole of one side of the link (figure 1.8; Plate 1) clearly indicated high-stress regions in the inside surface near the points of application of the load. Since the locations of these high-stress areas had been anticipated prior to testing, sections of the top loading link had been cut-away to provide line-of-sight access to these regions. Further scans were carried out to obtain improved spatial resolution and thus more accurate estimates of stresses in one of the high-stress regions by moving the scanning unit to a new viewing point which was more normal to the high-stress point. The apparent high stresses at the edges of the link are caused by significant motion normal to the viewing angle and should be ignored.

Another structure which underwent fatigue testing in relation to offshore applications is the large cast steel X-joint shown in figure 1.9. It was analysed in a client's structural testing laboratory using the NEL equipment. The amplitude and polarity of the stress distribution, both on the outside of the structure, and around the inside of the bore of the horizontal member were clearly revealed in the thermoelastic data. The overall stress distribution measured over the whole of one side of the entire structure can be seen in figure 1.10 (Plate 1). Four strain gauges and associated wiring are clearly visible around the saddle point on the outside surface of the joint. Such tests have indicated the technical and economic advantages of thermoelastic full-field stress measurements on large-scale structures, compared to the much more highly labour-intensive nature of strain gauge techniques. One of the very useful features of the thermoelastic technique was utilised in this series of tests. A scan of a large area was obtained in order to determine the highest-stress region on the horizontal member, then a 'zoom analysis' was carried out until the highest-stress point had been located within a few millimetres. The value and position of this high-stress point was confirmed from strain gauge readings. The stress patterns obtained using this technique offer additional insight into the behaviour of structures in a way that discrete-

**Figure 1.7** Marine chain link.

point methods just cannot match. The X-Node testing programme has been described in detail by Wood *et al* (1987).

A further example from the offshore industry illustrates the ability to resolve fine detail in regions where stresses are changing rapidly. Figure 1.11 shows a large tubular welded T-joint which was being fatigue tested under sinusoidal loading. The joint was loaded along the axis of the vertical member, applying bending to the horizontal member which was rigidly bolted by end flanges. Figure 1.12 shows the stress distribution measured over an area close to the weld. The details of individual weld runs show clearly in the stress pattern. A number of these joints have been analysed and in one of them a crack was detected. The crack was detected at the very early stages of its development, in the high-stress region at the toe of the weld. A series of scans performed at regular time intervals during the course of the test clearly showed the progressive crack growth. A second area of one of the T-joints was investigated in order to determine whether it would be possible to detect external surface stresses associated with internal stiffeners which were not visible on the smooth outer surface of the chord. The tests were again successful, revealing two bands of high stress corresponding to the geometry of the internal stiffeners.

The thermoelastic technique has also been applied to a number of structures relating to an on-going bio-engineering project at NEL. The technique was used as a means of identifying stress distributions in human

**Figure 1.9** Cast X-joint from an offshore platform.

bone by studying partial human femurs which were undergoing dynamic loading in a servo-hydraulic test machine. Tests to date show that it is possible to obtain stress patterns from fresh bone and that the emissivity from the cleaned bone is sufficiently high without resorting to the application of a high-emissivity paint (Duncan 1988). Various steel and titanium hip prostheses have also been tested as part of the same project. Current designs and materials used in total hip replacement prostheses have excellent mechanical properties and in themselves are unlikely to be the cause of primary failure, but loosening of the prosthetic stem in the fixation medium and the breakdown of the fixation medium itself may cause secondary failure of the prosthesis. The stress distribution associated with this condition is therefore of interest in order to determine the most likely site of failure. Figure 1.13 shows a hip prosthesis embedded in fixing cement ready for testing and figure 1.14 (Plate 2) displays its thermoelastic stress distribution. The region of highest stress on the prosthesis is seen to be in the area close to the surface of the fixation medium. The thermoelastic stress analysis was

**Figure 1.11** Tubular welded T-joint.

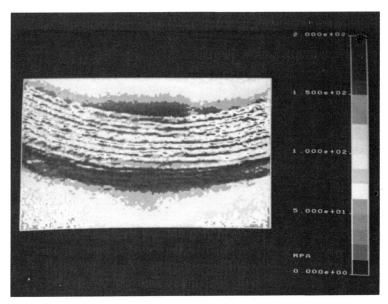

**Figure 1.12** Stress pattern of the T-joint weld of figure 1.11.

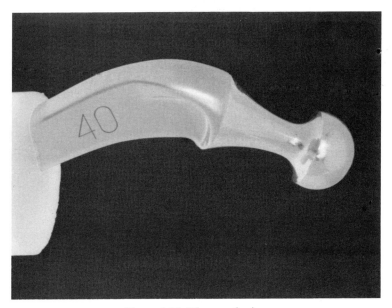

**Figure 1.13** Hip prosthesis.

performed during a fatigue test, and the point of failure was later found to be in the region close to the surface of the fixing cement.

As part of a study into the fatigue performance of wire ropes intended for offshore applications, an experimental stress analysis was performed on a short section of a 70 mm diameter, six-strand rope containing a total of 246 steel wires. A wire rope is manufactured by spinning a collection of wires into strands, several of which are then wound together to form a single rope. A SPATE area scan revealed distinctly the surface stress distribution in the strands and individual wires in that section of the rope. Despite the fact that the thermoelastic patterns are only a surface stress estimate, broken wires may appear as unstressed areas when they emerge at the surface. Thus, the thermoelastic technique shows potential for non-destructive testing (NDT) applications in such components.

Thermoelastic stress analysis was used in conjunction with a brittle lacquer/strain gauge experimental stress analysis of a commercial vehicle chassis cross-member in order to refine a finite element (FE) model of the component. The structure and the corresponding SPATE data were described in detail by Loader *et al* (1987). Stress concentrations around two bolt heads were revealed clearly in the thermoelastic stress distributions. The stress patterns measured on swaged and flat end-caps were also compared in order to ascertain the effectiveness of the stiffened configuration in reducing the stress concentrations around the bolt heads. Another area containing five puddle welds was analysed on the base of the end-cap. The thermoelastic analysis revealed that there were high stresses around just one puddle weld site, and this result was also obtained from a finite element model of the same component.

A prototype T4 railway bogie, manufactured by British Rail Engineering Ltd, was tested as part of an experimental programme to develop high-speed bogies, during which it was subjected to a variety of load configurations which simulated events predicted to occur in service. The thermoelastic stress patterns obtained during dynamic testing (Cummings and Harwood 1987) were found to be very useful in assessing the design of the bogie and confirming the accuracy of a finite element model of the structure. The bogie later achieved a world speed record during track testing.

An experimental stress analysis of an aluminium model of a con-rod was performed as part of a programme to evaluate the suitability of the component for automotive applications. The con-rod had previously been coated so that a photoelastic stress analysis could be carried out. It was, nevertheless, possible to measure the thermoelastic response on the coating. The use of the thermoelastic technique in conjunction with the reflection photoelastic technique allows the individual principal stresses to be separated, since the principal stress sum and difference may be estimated at every point on the surface of the component.

A fracture mechanics study, containing both theoretical and experimental contributions, has been carried out on two types of notched testpieces. The theoretical study of the crack tip stress field, performed by Pukas (1987), was made in order that the effect of higher-order terms and other influences upon stress intensity factor solutions could be determined. The measured stress pattern (figure 1.15; Plate 2) and the theoretical solution matched closely. Related work at Manchester University is described in Chapter 5. The results indicate that the thermoelastic technique has considerable potential for the future in this important application.

## 1.8 ACOUSTIC HOOD

Following unsuccessful attempts to use the NEL SPATE equipment to locate stress concentrations on an aircraft engine turbine blade operating at frequencies between 2 kHz and 8 kHz, it was realised that the signal-to-noise ratio was very poor due to interference from acoustic noise. The SPATE head had to be placed very close to the structure due to the confined space around it. Acoustic noise around the SPATE aperture was measured at 120 dBA, the noise emanating from close to the structure itself. These adverse conditions proved so severe that the equipment could not produce consistent data. A subsequent test under more controlled conditions using a stiffened plate vibrating at frequencies up to 1.3 kHz confirmed that the SPATE system is sensitive to acoustic noise: that where the noise is generated by the structure it correlates strongly with the reference signal and so is difficult to reject; and also that sensitivity to this type of noise is dependent on scanning mirror position. This latter point means that the normal practice of measuring an offset due to noise which can then be subtracted from the measured signal is not feasible. It was found that the effects of acoustic noise could be reduced by careful positioning of the SPATE head, the use of acoustic baffles and by viewing the structure via a mirror with physical dimensions small compared to the wavelength of the interfering sound. These approaches, however, do not offer a generally applicable solution to the problem. Therefore, it was decided to design an acoustic hood capable of substantial attenuation of acoustic noise. Note that the more flexible software and enhanced post-processing capabilities of the SPATE 9000 (see Chapter 3) make it more suitable than the 8000 system for noise subtraction of complex contributions on a point-by-point basis.

The body of the hood is constructed from sheet metal riveted to an aluminium frame. The sheet metal consists of two layers of 20 swg steel bonded by a layer of viscous plastic providing good damping. The hood is airtight but with provision for venting internal pressure generated by the evaporation of liquid nitrogen from the detector Dewar. Internal air spaces are kept to a minimum to reduce the possibility of acoustic resonance within the hood, and sound-absorbent foam is used where appropriate. Normally the only physical contact between the SPATE head and the hood will be via rubber isolators at the base and an optional securing screw passing through the hood to the tripod.

A window must of course be incorporated in the hood to allow the IR signals to reach the detector. Barium fluoride was initially considered for the window material since it has the advantage of being transparent in the visible range as well as much of the infra-red. However, this material was rejected due to its progressive attenuation of wavelengths above 10 $\mu$m. Therefore, germanium was considered to be more suitable, in a doped, monocrystalline form 8 mm thick with a high efficiency, anti-reflective coating on both surfaces. This gives minimal transmission losses over the 8 to 13 $\mu$m bandwidth together with acceptable mechanical properties. Rather than incorporating the window in the hood itself, it is inset into the aperture in the SPATE head and set at an angle of 25° to the optical axis, occupying approximately the position of the polythene pellicle in the SPATE head. This reduces the internal free air space and gives a wide angle of view for minimum window area, so reducing cost and increasing stiffness. The angled window is expected to reflect sound better and avoids the narcissus effect where the detector sees its own cold reflection, possibly modulated by vibration of the reflecting surface.

Germanium has the disadvantage that it is opaque to visible light. Therefore, in order that the optical channel of the SPATE system may be used for optical alignment and the setting of markers, provision has been made for the substitution of a glass window. The angle and thickness of this glass window may be chosen to give the same displacement of the viewed image due to refraction as when using the germanium window, so allowing accurate placing of markers. The top two-thirds of the hood must be removed to give access to the telescope of the optical channel and the IR channel focusing ring. This leaves the base of the hood with cable seals and window aperture undisturbed. A plug is provided for recharging with liquid nitrogen, with the hood otherwise sealed.

## 1.9 CONCLUSIONS AND FUTURE DEVELOPMENTS

Over recent years the number of SPATE users has steadily expanded throughout the developed world. Their experience has demonstrated that the thermoelastic technique is a rapid and convenient method of performing full-field experimental stress analysis on dynamically loaded structures. The stress estimates are produced in a colour-coded form which is very easy to interpret and the technique has been found to be suited to a broad range of troubleshooting and design-appraisal applications. The technique also requires very little surface preparation and has been applied to a wide range of structural materials; at NEL thermoelastic response has been measured on several steels, aluminium, titanium, brass, copper, Nimonic 90, zirconium, Perspex, Tufnol, brick, concrete, fresh bone, and several ceramics, fibre-reinforced plastics and woods. The non-contacting form of measurement and the ability to use the equipment at a sufficient stand-off distance to examine large areas of a structure and then zoom in by physically moving the head closer to the regions of main interest have been found to be particularly useful features. The practical exploitation of the thermoelastic effect clearly ranks with recent laser-technology advances, such as moiré or holography, as a key development in experimental stress analysis.

Thermoelastic stress analysis has great potential in engineering design, particularly for fatigue conditions, for which a knowledge of the location of stress concentrations is vital, and where failures in service can be cata-strophic. Nevertheless, since the thermoelastic output is insensitive to shear stresses, it is more realistic for many applications to use such equipment in conjunction with other experimental or analytical methods, rather than reliance on thermoelastic data alone to locate stress concentrations. The technology complements existing computer-based analysis methods by pro-viding prompt experimental feedback to validate these theoretical tech-niques, thus reducing the lead time in the design cycle. The use of such technology as a design aid or for troubleshooting in service is an invaluable new addition to the armoury of experimental techniques which, if exploited fully, is bound to enhance product quality in many sectors of industry.

Thermoelastic stress analysis is a new technology which is developing rapidly. It can produce the experimental data which are the basis for cost-effective and reliable design to give optimum structural efficiency. SPATE is particularly effective when used in conjunction with the FE technique to provide the experimental data which may confirm theoretical solutions obtained from idealisations of the structure. Such experimental confirmation gives confidence in the accuracy of the FE model. Thermoelastic stress patterns may be used to optimise the model by indicating high- and low-stress gradients where a finer or coarser mesh size would be more appro-priate. The data from the FE analysis and the experimental stress patterns can be matched in the form of the sum of the principal stresses to provide a method of validating the model. The distributions of the principal stress vectors may then be obtained from the verified model which of course can also then be used to calculate deformations or the von Mises effective stresses for failure prediction under multi-axial conditions.

Thus, although the direct value of SPATE stress data is limited because it is in the form of the sum of the principal stresses, the thermoelastic technique is well suited to integration with theoretical analysis in order to extract parameters more relevant to engineers involved in design for strength, durability and structural efficiency. This methodology is applicable to both static and modal data and offers considerable potential for savings in cost and time if incorporated in the design procedures for engineering components and structures.

At this stage it is not expected that the thermoelastic technique will replace other experimental techniques, but rather that it will complement them and expand into new areas which are not presently available to stress analysts. The most obvious use of the SPATE system is in identifying high-stress regions in prototypes. Several prototype designs may be manufactured and tested in turn, making the selection of the most appropriate design a fairly simple task. A more successful result may thus be achieved by repeating the procedure until a satisfactory design has evolved. Product development time may therefore be reduced if a rapid experimental technique such as thermoelastic stress analysis is used rather than more time-consuming methods.

Areas which are currently being developed or may be developed in the future are listed below:

(i) The use of multiple detectors (e.g. staring arrays) and the improvement of scanning and signal processing technology to speed up scan times allow transient load conditions to be analysed and make the technique more suitable for quality assurance applications;

(ii) the extension of analytical techniques to allow principal stress vectors to be separated for a wider range of structures;

(iii) investigations into the feasibility of estimating residual stresses from second-order thermoelastic effects or from thermal effects using the TERSA technique (Mountain and Cooper 1989);

(iv) more developments in the area of target motion compensation techniques;

(v) improved techniques for high-frequency conditions, including acoustic noise suppression and the selection of appropriate coatings;

(vi) development of analytical methods for the interpretation of data acquired from anisotropic materials; and

(vii) development of a lens system to provide finer resolution in areas of rapidly changing stress.

## ACKNOWLEDGMENT

This chapter is published by permission of the Chief Executive, National Engineering Laboratory, Department of Trade and Industry. It is Crown copyright.

# 2

# Thermal Analysis of Deformation Mechanisms Using a Contacting Technique

## M G Beghi

## 2.1 INTRODUCTION

In solids under stress, different types of deformation processes occur. These vary according to the nature of the solid, its microscopic structure, and physical conditions such as the temperature or the stress level. Different processes correspond to different microscopic mechanisms, and exhibit different macroscopic behaviours. The analysis of these behaviours can be undertaken, at a macroscopic level, in terms of mechanical quantities, namely stress fields and strain fields, and more comprehensively by also considering thermodynamic quantities like temperature fields.

This chapter focuses on the exploitation of temperature measurements in the analysis of the deformation processes, mainly of metals. The different deformation regimes are reviewed in section 2.2, and the possibility of developing a thermodynamic theory of deformation is analysed. Several temperature measurement systems are considered in section 2.3, and their potential for this application is assessed. Contacting techniques turn out to be more suitable. Several results, obtained in both monotonic and cyclic tests, are then reviewed in sections 2.4 and 2.5.

## 2.2 THERMOELASTICITY AND THERMOPLASTICITY

### 2.2.1 Classification of experimental behaviour

The main features of the typical behaviour of solids undergoing deformation

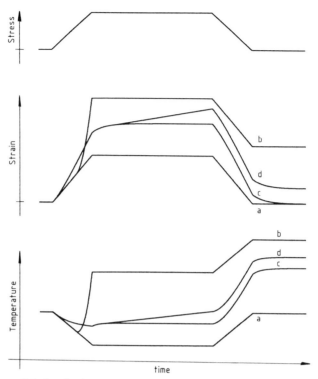

**Figure 2.1** Qualitative representation of the strain and temperature variation due to an imposed trapezoidal load for (*a*) an elastic solid, (*b*) an elastic–plastic solid, (*c*) a viscoelastic solid, and (*d*) a 'viscous' solid.

are briefly summarised in this section. For simplicity, uniaxial tension conditions are considered, in which the deformation state can be represented by the axial stress $\sigma$ and the axial strain $\varepsilon$, and the stress and strain fields are homogeneous. The stress is assumed to be imposed, and is considered as the excitation, or the input; the strain being regarded as the response, or the output. Parallel considerations can be developed for the dual case in which the strain in imposed. Throughout section 2.2.1, reference is made to a representative loading history formed by a loading stage, in which the stress is raised at constant rate from zero to a maximum tensile stress. This is followed by a hold time at maximum stress and an unloading phase, in which the stress is decreased to zero at the same rate, and a final hold time at zero stress (figure 2.1).

The temperature variations induced by the deformation must also be considered as a response. For a homogeneous stress and strain field, the temperature field ($T$) is also homogeneous provided that the boundary conditions are adiabatic. If instead heat is exchanged with the environment, heat conduction determines a space- and time-dependent temperature field. Whenever the stress and strain fields are non-homogeneous, the temperature field is also non-homogeneous and heat conduction occurs within the solid. Heat conduction is not significantly affected by deformation processes. In the present discussion we can neglect any such effect and assume a perfectly constant thermal conductivity, $k$. It must be remembered that heat conduction is a linear time-dependent phenomenon which is thermodynamically reversible only in the idealised case of vanishing temperature gradients, i.e. of quasi-static processes. In real cases it is always irreversible, although this irreversibility has small practical relevance, as the temperature variations and temperature gradients due to deformations are usually small.

The simplest mechanical behaviour is linear elastic, defined by a simple proportionality between stress and strain: $\sigma = E\varepsilon$, where $E$ is Young's modulus. Under adiabatic conditions, the temperature is also proportional to the stress, via a negative thermoelastic constant (see sections 1.5.1 and 2.2.2). Cooling occurs at a constant rate in the loading phase, and a symmetric heating occurs in the unloading phase. Nowick and Berry (1972) note that elasticity implies: (i) the existence of a unique equilibrium strain corresponding to each stress level; (ii) the instantaneous achievement of the equilibrium response; and (iii) linearity of the response. Linearity takes into account the time dependence, and means that the response (strain or temperature history) to the sum of two excitations (load histories) is the sum of the responses which are obtained when each of the two excitations is applied alone.

The existence of a unique equilibrium strain corresponding to any stress level implies the complete recoverability of strains. The mechanical energy per unit volume, $Q_m$, which is expended in the loading phase can be completely recovered in the unloading phase. Nevertheless, because of the possibility of thermal gradients occurring, a strict thermodynamic reversibility is obtained only in perfectly adiabatic or perfectly isothermal conditions. This irreversibility also affects mechanical reversibility by quantities which are, at most, of the order of the relative difference between the adiabatic elastic modulus and the isothermal one. Non-linear elasticity can also be considered, to which the above considerations about reversibility also apply. Nevertheless, this type of behaviour is encountered relatively infrequently, and will not be considered further.

The ensemble of non-elastic regimes is often called 'inelasticity'. Standard behaviours can be considered, which are idealisations of experimentally observed behaviours, and can be classified by making reference to the three characteristics of elastic behaviour mentioned previously. The response of a real solid to an applied stress is often a combination of an elastic component with one or more standard inelastic components.

If the response is still instantaneous, but a unique equilibrium strain cannot be found corresponding to each stress level, the behaviour is plastic.

Most metals exhibit a transition from elastic to plastic behaviour when the stress exceeds the yield stress. The stress–strain trajectory deviates from linearity, the strain growing to substantially higher values than those corresponding to elastic strains. Being a threshold phenomenon, plasticity is intrinsically non-linear. In uniaxial tension conditions the total strain $\varepsilon$ can be uniquely decomposed into the sum of an elastic strain $\varepsilon_e$ and a plastic strain $\varepsilon_p$, where $\varepsilon_e = \sigma/E$ as in the elastic case. The plastic strain $\varepsilon_p$ is a permanent set which remains constant during the hold times and also in the unloading phase, in which the elastic strain is recovered. Thus, any given stress level, such as for instance the initial and final zero stress, can correspond to any value of $\varepsilon_p$.

Only the elastic part $Q_e$ of the mechanical energy $Q_m$ expended in the loading phase is recovered in the unloading phase. The plastic part $Q_p = Q_m - Q_e$ is partly converted into heat and partly stored in the solid (stored energy of cold work). The energy converted into heat produces a temperature rise (thermoplastic effect), the amount of which depends on the experimental conditions, and can be significantly larger than that due to the thermoelastic effect. Although the stored energy cannot be retrieved mechanically, it may be recovered by thermal annealing; it corresponds to a permanent modification of the microscopic state of the solid. The plastic deformation can be recovered by imposing an opposite plastic strain. A tension/compression cycle can be performed, at the end of which both the stress and the strain have returned to the initial value. Nevertheless, energy conversion and storage occur during both branches of the cycle, so that part of the mechanical energy expended during the cycle is permanently stored in the solid as internal energy. Although the stress and the strain have returned to the initial values, the initial state of the solid is not fully recovered. Both the energy conversion and the energy storage are intrinsically irreversible processes. The plastic deformation is irreversible irrespective of the rate at which it occurs, and no reversible path exists which generates a plastic deformation (Bridgman 1950). Furthermore, once plastic deformation has occurred, the initial state of the solid cannot be recovered, except by processes like thermal treatments.

Considering next the regimes in which the response is not instantaneous, a very fast loading stage is often adopted in order to discriminate the non-instantaneous part of the response from the instantaneous one. Then the time-dependent response occurs entirely during the hold time at maximum stress. Namely, strain increases at constant stress, a phenomenon called creep, and it decreases during the hold time at zero stress after the unloading phase (this decrease is sometimes called elastic after-effect). Stress relaxation at constant strain is another manifestation of the same behaviour (Christensen 1971).

Two standard behaviours can be distinguished according to whether or not a unique equilibrium strain exists corresponding to each stress level. If a unique equilibrium strain exists, creep occurs at a continuously decreasing rate, asymptotically approaching a finite strain. Upon unloading, part of the strain is instantaneously recovered, the rest of it being completely recovered by the elastic after-effect. The initial state of equilibrium at zero stress is eventually reached (see figure 2.1). This type of behaviour is exhibited by viscoelastic solids. If a unique equilibrium strain does not exist, creep eventually reaches a constant strain rate, the strain increasing until fracture occurs. Upon unloading, the strain achieved in the constant-strain rate regime is not recovered. This type of behaviour is typical of viscoelastic liquids, and will be called here 'viscous'. Viscoelastic and viscous solids keep memory of their preceding states; the memory is fading in the viscoelastic case, and permanent in the viscous case. Both viscoelastic and viscous behaviours can be linear or non-linear.

The time dependence of the response indicates that the state which is instantaneously reached is not an equilibrium one; the deformation can be recoverable, but it is not reversible. For any finite loading rate, the unloading stress–strain trajectory is different from the loading one, and the expended mechanical energy $Q_m$ is not completely recovered. In the viscoelastic case, the state which is eventually reached after a loading/unloading cycle is a unique equilibrium state, identical to the initial one; the memory of the completed cycle fades completely. Accordingly, energy is not permanently stored; the part of the expended energy which is not recovered is completely dissipated into heat. The amount of dissipated energy depends on the loading rate. If the loading stage tends to a vanishing rate (quasi-static limit), equilibrium tends to be achieved for any infinitesimal load increment, and the deformation tends to reversibility. In contrast to the plastic and the viscous regime, reversible paths can thus be identified connecting any pair of possible states. In the viscous case the deformation is thermodynamically irreversible irrespective of the rate at which it occurs.

Most metals have an essentially elastic–plastic behaviour, a small component of their response being viscoelastic. However, in high-temperature creep a secondary stage is usually encountered, in which a permanent strain is achieved at constant rate under constant stress, thus behaving in a viscous fashion. Viscoelastic deformations and high-temperature creep usually occur at low to very low strain rates, thus involving low rates of heat generation. Consequently, while temperature variations due to the thermoelastic and thermoplastic effects are measurable and can be exploited to study these deformation regimes in metals, temperature variations due to other deformation regimes are significantly smaller and are difficult to measure. In polymers, on the other hand, viscoelasticity can be important. Thermal effects can be exploited to study the deformation phenomena, and must always be taken into account when analysing temperature measurements on deformed polymers.

In metals, at a microscopic level, elastic deformations correspond to variations of the inter-atomic distances, the crystalline lattice remaining unchanged. On the other hand, plastic deformations correspond to the motion and multiplication of line defects of the lattice, i.e. the dislocations (see e.g. Friedel 1964, Nabarro 1967, Hirth and Lothe 1968). The motion of single dislocation segments is hindered by various mechanisms which result in a conversion of mechanical energy into thermal energy. In a real crystal, dislocations can be blocked by pinning points like intersections with other dislocations and impurities, and by obstacles like precipitates and grain boundaries. Dislocations can be unpinned, if the stress exceeds appropriate thresholds, and can then move until they meet another obstacle. Dislocation sources can be activated above characteristic stress thresholds. The irreversibility of plastic deformation is determined by the nature of its microscopic mechanism. Once a dislocation has moved, it does not move back under a stress removal, but requires a full stress reversal. But even though a single dislocation segment can reverse its motion under a stress reversal, a complex dislocation network will not simply reverse its evolution. For instance, once dislocations have been created and have moved away from their source, under a stress reversal they do not necessarily move back to the source and disappear. Similarly, if dislocations have disappeared, because they have annihilated or reached the surface, they do not reappear by reversing the stress. The energy which is stored during plastic deformation is stored as a microscopically elastic strain energy of the complex microscopic deformation field due to the dislocation network.

## 2.2.2 Theoretical approach

In this section a thermodynamic approach to the description of the previously mentioned behaviours is outlined. The applicability of thermodynamic concepts is discussed first. The theory of the thermoelastic effect is then recalled, as well as the theory of the thermo-viscoelastic effect, and an empirical treatment of the thermoplastic effect is discussed. Other deformation regimes are not considered, because, as mentioned above, temperature measurements are not viable in their analysis.

As for any thermodynamic theory, the development of a theory of the deformation of solids in the framework of non-equilibrium thermodynamics is only possible if the thermodynamic state of the solid can be precisely defined. Moreover, the entropy variation between any possible state must be capable of being evaluated, so that local values of quantities like internal energy, entropy and free energy can be unambiguously defined. The physical state of a small, but still macroscopic, volume of a solid is defined by local values of macroscopic variables, such as the stress tensor and the temperature, and by its microstructure. The generic term 'microstructure' indicates a whole spectrum of features. In the case of crystalline materials, the microstructural state is determined by the distributions of point defects (vacancies and interstitials), by the dislocation network, and possibly by mesoscopic defects like grain boundaries or precipitates. In the case of polymers other quantities are relevant, e.g. the amount of crystallinity. In some cases a small set of internal state variables can be identified, which completely specifies some features of the microstructure. This is often the case for point defects. In other cases, e.g. dislocation networks, a description of the microstructure in terms of a limited number of variables is not possible, and often it is also the case that a complete observation is impossible.

A theory of the deformation of solids does not need to give a full account of the possible microstructural states, which are the outcome of the production process of the material. It must only consider the features that can be modified by the deformation process, neglecting all the features which remain unchanged. A distinction thus arises between the deformation regimes involving microstructural modifications which can be described by a small set of internal variables, and those involving modifications for which such a description is not possible. In the first case the thermodynamic state of the solid is unambiguously specified by a limited set of macroscopic and internal variables. In the second case such a definition is not possible.

The elastic case is the simplest one, in which the microstructure is not altered at all. The state of the system is completely specified by the macroscopic variables. As already mentioned, in the elastic regime reversible paths are easily identified between any possible state. Entropy differences can thus be evaluated by Clausius' theorem. A thermodynamic theory can therefore be developed for this regime. Its main results are outlined here. The reader should refer to Chapter 7 for a more complete coverage. A local entropy balance equation can be established which relates the entropy variation to entropy fluxes and entropy production. In the elastic regime, the entropy variation is the sum of a vibrational and a configurational entropy variation. These are functions only of the temperature and volume ($\delta\varepsilon$) variations respectively (Biot 1956, Landau and Lifshitz 1970). Entropy fluxes are proportional, via a ($1/T$) factor, to heat fluxes. Entropy productions are due to irreversible phenomena. The entropy production due to heat conduction along a non-vanishing gradient (the only possible irreversible phenomenon in the elastic case) is well known (de Groot and Mazur 1962). The entropy balance equation can be stated in the form of a Fourier equation for the temperature field $T$, of the form (Biot 1956)

$$\rho C_v \frac{\partial T}{\partial t} - k \nabla^2 T = P_{te} \qquad (2.1)$$

where $\rho$ is the density ($kg\,m^{-3}$), $C_v$ is the specific heat at constant volume ($J\,kg^{-1}\,K^{-1}$), and $k$ is the thermal conductivity ($W\,m^{-1}\,K^{-1}$). The term $P_{te}$, the 'thermoelastic effective source', has an entropic origin and plays the role of a heat source (its dimensions are $W\,m^{-3}$). It should, however, be remembered that it is not a proper heat source, because it does not correspond to a generation of thermal energy, but enters the equation because of the dependence of the entropy on the volume variation. $P_{te}$ can be expressed as

$$P_{te} = -3\alpha K T \frac{\partial(\delta\varepsilon)}{\partial t} = -\rho C_v \gamma T \frac{(\delta\varepsilon)}{t} \tag{2.2}$$

where $\alpha$ is the coefficient of linear thermal expansion ($\mathrm{K}^{-1}$), $K$ is the bulk modulus (Pa), and $\gamma = 3\alpha K/\rho C_v$ is the Grüneisen parameter (Wallace 1972). If the strain is homogeneous and the boundary conditions are adiabatic, the temperature field is homogeneous. Equation (2.1) can then be immediately integrated, obtaining equation (1.2). This specific case can clarify the entropic interpretation of the thermoelastic effect. If the temperature field is homogeneous, both the entropy flux and the entropy production are zero. The transformation is reversible and adiabatic, such that entropy remains constant. If the volume increases, the configurational entropy increases, and the vibrational entropy (therefore the temperature) must decrease by the same quantity to conserve the total entropy.

Equation (2.1) governs the temperature field in a solid undergoing elastic deformations. In particular it governs the attenuation of the thermoelastic signal by heat conduction, which is discussed in section 1.1.2. If heat sources are present which are not due to the deformation itself, they introduce additional terms on the right-hand side of equation (2.1). In uniaxial tension conditions the volume variation $\delta\varepsilon$ is given by $\delta\varepsilon = \varepsilon(1 - 2v)$, where $v$ is Poisson's ratio.

Considering next the viscoelastic case, it is often possible to identify a small set of internal variables that define the microstructural modifications occurring during the viscoelastic deformations (e.g. anelastic relaxations of point defects). The state of the system can therefore be completely specified. It has already been noted that, in the viscoelastic regime, reversible quasi-static paths can be identified between any possible state, so that entropy differences can be evaluated. A thermodynamic theory can therefore be developed also for the viscoelastic regime. In particular, in the linear viscoelastic case, linear relationships can be found between the rate of change of the internal variables and the stress. A basis is thus found for the unique correspondence between equilibrium states and stress levels (Nowick and Berry 1972). In these cases an expression can be found for the entropy production term. In the language of non-equilibrium thermodynamics, a linear relationship exists between thermodynamic forces and thermodynamic fluxes (de Groot and Mazur 1962). Without mentioning the details, which depend on the specific viscoelastic process, an entropy balance equation can be established also in this case. The equation can be expressed in the form of (2.1), where an additional term $P_v$ appears on the right-hand side of the equation. As for $P_{te}$,

$P_v$ plays the role of a heat source in units of $\text{W m}^{-3}$. In contrast to $P_{te}$, it is a proper heat source, because it is directly related to the dissipation of mechanical energy by the viscoelastic solid. Note that it also corresponds precisely to an entropy production term.

Considering finally the plastic case, it has already been noted that plastic deformations correspond to changes in the dislocation network within the solid, and that these cannot be specified by a set of parameters (quantities such as dislocation densities are not sufficient). Thus the thermodynamic state of a solid undergoing plastic deformations cannot be unambiguously defined. Furthermore, if a state is reached by a plastic deformation, no reversible path can be identified which connects it to the initial state; the entropy of a generic state cannot be evaluated. Therefore a thermodynamic theory of plastic deformation cannot be developed. An empirical treatment must instead be adopted, involving energetic but not entropic quantities, since these cannot have a quantitative expression.

It has already been noted that if the mechanical energy per unit volume $Q_m$ is expended in the loading phase, and plastic flow occurs, only the elastic part of it, $Q_e$, is recovered in the unloading phase, the plastic part $Q_p$ being irrecoverable. A part of $Q_p$, called $Q_d$, is immediately dissipated into heat; the complement $Q_i = Q_p - Q_d$ is stored in the solid. The corresponding rates (powers per unit volume) are $P_m$, $P_e$, $P_p$, $P_d$ and $P_i$, respectively. In the uniaxial tension case (i.e. $\varepsilon = \varepsilon_e + \varepsilon_p$), $P_m$, $P_e$ and $P_p$ are given by $P_m = \sigma \dot{\varepsilon}$, $P_e = \sigma \dot{\varepsilon}_e$ and $P_p = \sigma \dot{\varepsilon}_p$, where the dot indicates the time derivative. A dimensionless fraction $f = P_d / P_p$, known as the 'dissipation function' (Beghi *et al* 1986), can be defined to express the fraction of the expended plastic power which is immediately converted into heat. The dissipation function depends on the microstructure of the material and of the deformation conditions such as temperature, strain rate, stress or strain level or triaxiality. The term $P_d$ is a proper heat source, which must be added on the right-hand side of equation (2.1). It is the outcome of an irreversible phenomenon, but it cannot be given a precise meaning in terms of entropy production. Equation (2.1) becomes, for an elastic–plastic solid,

$$\rho C_v \frac{\partial T}{\partial t} - k \nabla^2 T = -\rho C_v \gamma T \frac{\partial(\delta \varepsilon)}{\partial t} + P_d. \tag{2.3}$$

As the temperature variations due to the deformation are usually small, the temperature, $T$, appearing on the right-hand side in the expression of the thermoelastic effective source can be substituted by a constant temperature $T_0$, typically the initial temperature of the solid; the equation is thus linearised. Introducing the variable $\theta = T - T_0$ and the thermal diffusivity $\chi = k / \rho C_v$, equation (2.3) takes the form (Beghi *et al* 1986)

$$\frac{\partial \theta}{\partial t} - \chi \nabla^2 \theta = -\gamma T_0 \frac{\partial(\delta \varepsilon)}{\partial t} + \frac{P_d}{\rho C_v}. \tag{2.4}$$

## 2.3 MEASUREMENT AND ANALYSIS OF THE THERMO–MECHANICAL BEHAVIOUR OF SOLIDS

### 2.3.1 Temperature measurements

Temperature variations induced by the deformation processes discussed in section 2.2 can be measured at the surface of solids undergoing deformation. Several measurement techniques are available, the main distinction being between contacting and non-contacting techniques. The suitability of the available techniques for different types of measurement is assessed in this section.

Only a few materials, e.g. concrete, allow the preparation of specimens in the bulk of which temperature sensors are embedded. In the great majority of materials only the surface temperature is accessible. The significance of surface temperature must therefore be assessed when the stress at the surface does not coincide with the stress in the bulk. For instance, in the bulk of thick sections plane strain conditions can prevail, while at the surface plane stress conditions prevail. The considerations mentioned in sections 1.1.2 and 2.3.2, and presented more extensively in Chapter 4, apply. If the loading frequency is high enough to guarantee adiabatic conditions, the surface temperature depends on the surface strains only, while at lower frequencies the surface temperature also depends on the strain in a surface layer of non-vanishing thickness. Heat exchanges with the environment must also be considered, especially in the case of specimens with low conductivity, and at low loading frequencies. In such cases surface temperature fluctuations due to normal air currents can be significant. A simple box enclosing the specimen is often sufficient to provide shielding.

Contacting and non-contacting techniques are available to measure surface temperature variations. Contacting techniques include thermocouples, resistance thermometers or resistance temperature detectors (RTDs), and thermistors, while non-contacting techniques are based on IR sensors. A thermocouple is the junction of two different metals, at which an electromotive force is developed, which is a function of the temperature of the joint. RTDs and thermistors are metallic and semiconductor sensors, respectively, the resistances of which are a function of the temperature. A measurement system based on a thermocouple essentially amplifies the voltage generated by the thermocouple itself, while a measurement system based on one of the resistive sensors measures the resistance by a standard configuration, such as a bridge. IR detection, and the SPATE system based on it, are described in section 1.3, and in references therein.

Thermocouples exist in several standardised types, which allow different temperature ranges and different resolutions. The sensitivity of the J and K types (the common choices around ambient temperature) range between 40 and 60 $\mu$V K$^{-1}$. Resistive sensors are characterised by the temperature coefficient (TC), the ratio of the relative resistance variation ($\Delta R/R$) to the

# Plate 1

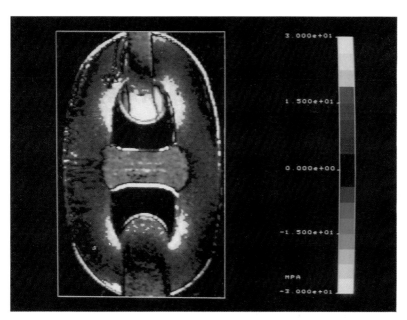

**Figure 1.8** Stress pattern from chain link

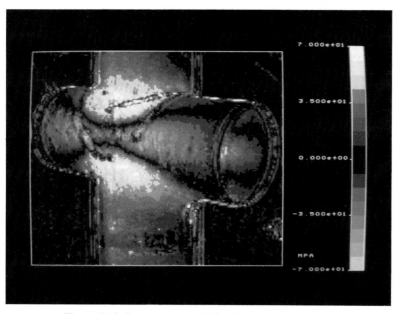

**Figure 1.10** Stress pattern of the X-joint of figure 1.9

# Plate 2

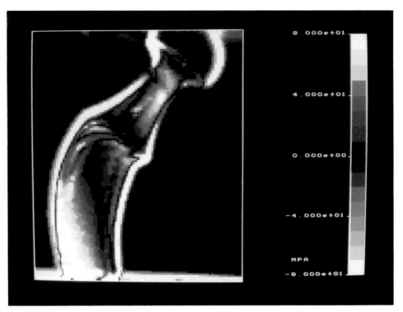

**Figure 1.14** Stress pattern from hip prosthesis of figure 1.13

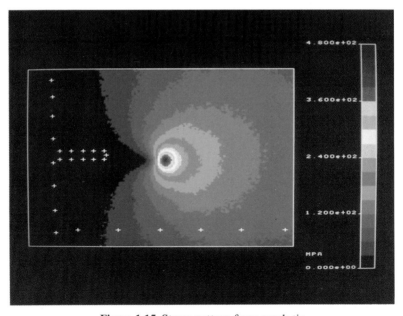

**Figure 1.15** Stress pattern from crack tip

# Plate 3

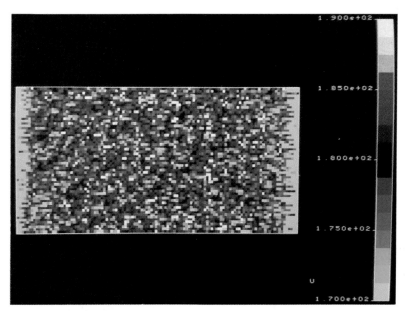

**Figure 7.5** SPATE scan of the originally straight specimen

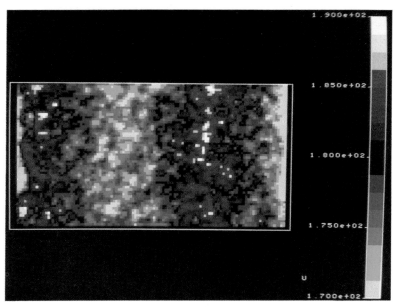

**Figure 7.6** SPATE scan of the straightened specimen

# Plate 4

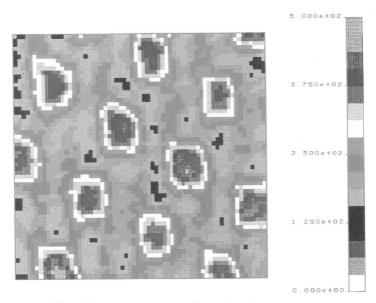

**Figure 8.5** SPATE scan of a carbon-fibre cloth composite

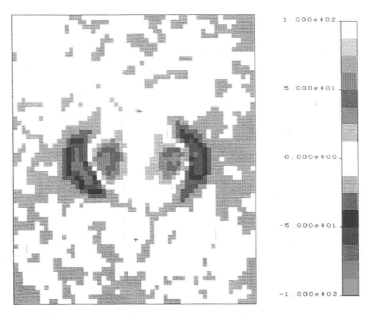

**Figure 8.9** SPATE scan of an impacted carbon-fibre composite laminate

temperature variation $\Delta T$. Platinum resistance thermometers are the most popular type of RTD. Their characteristic is practically linear over temperature variations of some tens of degrees. Common types have a resistance of 100 Ω at 0°C, with a temperature coefficient around $0.004\,K^{-1}$. Both positive temperature coefficient (PTC) and negative temperature coefficient (NTC) thermistors exist, according to the type of doped semiconductor; the resistance increases (PTC) or decreases (NTC) with temperature. The temperature–resistance characteristic is typically exponential. Common types have a resistance of a few kilohms to some tens of kilohms at ambient temperature, with a temperature coefficient of around $0.04\,K^{-1}$. Thus thermistors have higher temperature coefficients and therefore a greater sensitivity than thermo-resistances. On the other hand, the characteristics of RTDs are more easily reproducible, and are standardised. Platinum resistance thermometers can be used as reference sensors to calibrate thermistors. Sensors of the three types exist in a wide variety of dimensions, mountings, and costs.

The mounting of a contacting sensor must insure the lowest thermal resistance between the specimen surface and the sensor itself, and the lowest thermal capacity of the sensor assembly. These two parameters have little influence on the static response of the sensor, but they are the major factors determining the dynamic response of the assembly if small sensors having an intrinsic fast response are adopted. The sensors can be glued to the surface, or simply kept in mechanical contact with it by some support, the thermal contact being insured by e.g. silicone grease. Thermocouples can also be soldered to metallic surfaces. In the case of resistive sensors, electrical insulation from metallic surfaces must be insured by a coating of the sensor itself or by an appropriate mounting. In the case of non-contacting measurements the state of the surface, in particular its emissivity, is important (cf. section 1.3).

In the assessment of the viability of a type of sensor for a given measurement, the factors to be considered are the sensitivity of the sensor, its response time, the perturbation of the temperature field due to the sensor itself, and the operational flexibility. The sensitivity and stability of the sensors must be evaluated, keeping in mind the peculiarities of the signals to be measured. The temperature signals induced by deformations are small variations around an average temperature, which at most fluctuates slowly. The measurement of the average temperature is only needed to insure uniformity between measurement and calibration conditions. Temperature changes can modify the mechanical properties and the conductivity of the material, and the strength of the thermoelastic effect, which is proportional to the absolute temperature. As these effects are small, accuracies of the order of 1 K in the measurement of the average temperature are more than acceptable. The proper signal consists instead of variations which are of the order of 0.1 K in the thermoelastic case, and possibly larger in the thermoplastic case. A resolution of the order of 0.001 K is therefore needed. An exception to this occurs during high-frequency tension–compression cycling in the elastic–plastic regime, where temperature increases of several tens of degrees can be reached.

The resolution is essentially limited by the signal-to-noise ratio. The intrinsic noise of IR sensors is comparatively high, and it prevents the achievement of the indicated resolution in single monotonic measurement. However, filtering techniques based on the correlation with the load signal have been developed (see section 1.1.2 and Chapters 3 and 6), which allow the component of the temperature signal which is linearly correlated with the load to be extracted, and thus achieve the required resolution on that component. These techniques are most efficient with cyclic signals. The intrinsic noise of thermocouples is smaller, and allows the achievement of the above resolution with special electronic chains. Resistive sensors, and particularly thermistors, also have low intrinsic noise, and reach greater sensitivities. A 30 kΩ sensor, working under a constant current of 20 $\mu$A (which insures a very low power dissipation, see below) has a sensitivity above 20 mV K$^{-1}$, i.e. more than two orders of magnitude higher than that of a thermocouple. Thermistors can thus reach, on monotonic transient signals, resolutions of 0.001 K and even better, without very stringent requirements on the electronic chain. The noise which affects thermistor measurements is in fact mainly a measurement of true temperature fluctuations, as is shown by its sensitivity to the degree of shielding from air motion. The resolution which can be reached by thermistors without specific filtering is thus determined by the experimental configuration more than by the sensor itself. Resolutions better than 0.001 K can be reached by a careful calibration of the sensors and by a proper shielding, but without the adoption of a vacuum chamber.

The response of IR sensors is intrinsically very fast. The response of contacting sensors is much slower, and is determined by the type of mounting as well as by the sensor itself. The simple representation of the sensor assembly as a lumped heat capacity (the sensor) connected to a heat reservoir (the specimen) by a thermal resistance is often acceptable. This corresponds to a transfer function of the specimen temperature to the sensor temperature characterised by a single pole, i.e. a low-pass system with a cut-off frequency determined by the ratio of the thermal resistance to the heat capacity. Higher cut-off frequencies can be reached by the choice of small sensors and a glue (or a contacting grease) having a good thermal conductivity, and by carefully minimising the overall dimensions, i.e. the total thermal capacity of the assembly. Very small sensors are available; in particular, bead thermistors are available with diameters of the order of 0.3 mm and where electrical contacts are insured by thin (0.03 mm) wires. Foil-type RTDs are also available with a thickness of approximately 0.05 mm. Nevertheless, even with these small sensors, cut-off frequencies reach at most 1 Hz (Beghi *et al* 1987a); cut-off frequencies of a few hertz can be reached with thermocouples formed by very thin wires.

Considering then the temperature perturbation due to the measurement itself, the non-contacting technique has a substantial advantage. Contacting sensors can perturb the temperature they measure, by their finite thermal capacity and possibly by heat dissipation. The small sensors previously mentioned have, however, a very small thermal capacity. In particular, at frequencies below the mentioned cut-off frequencies, the sensor effectively exchanges heat with a specimen zone of at least a few millimetres, i.e. a zone whose thermal capacity is much larger than that of the sensor itself. The perturbation due to the thermal capacity of the sensor is thus negligible. Thermocouples have a completely negligible dissipation, since a voltage can be measured with virtually null currents, while resistive sensors are intrinsically dissipative, and introduce a self-heating. Only variations of the self-heating are harmful, however, because a constant component only adds a systematic error in the measurement of the average temperature. A compromise must be found between higher sensitivities requiring higher electric currents, and lower dissipations requiring lower currents. Measurement currents can generally be adopted, as in the previous example of a 20 $\mu$A current in a 30 k$\Omega$ thermistor, by which the dissipated power is of the order of 0.01 mW, while the mentioned resolution of 0.001 K is fully achieved. As is easily shown, with such dissipated powers spurious contributions due to variations of the self-heating are much smaller than other sources of uncertainty. Care must, however, be exercised when some specific sensitivity requirement imposes higher currents, when the measurement is performed on thin specimens (typically when the thickness is of the order of 1 mm), or when a poorly conductive material is used in which heat conduction to the sensor effectively occurs only from a smaller area of the specimen.

The calibration of an IR sensor can be done by comparison with a contacting sensor. Nevertheless, in thermoelastic stress analysis applications, a proper temperature calibration is not required; a direct stress calibration being straightforwardly obtained by performing a measurement on a specimen of the same material subjected to a known stress. The characteristics of thermocouples and of metal resistance thermometers are standardised, while thermistors can be purchased in selected types with characteristics which fall within a given tolerance from a given curve. To achieve a higher accuracy they can be individually calibrated *in situ* by comparison with a metal resistance thermometer, mounted in such a way as to guarantee a very good thermal equilibrium between the RTD itself and the thermistors.

As far as the operational flexibility is concerned, the IR sensor needs a liquid nitrogen supply, but it requires only an overall preparation of the surface, while contacting sensors must be individually fixed to the surface, in a way which may or may not allow re-use. Furthermore, as already mentioned, the best performances of contacting sensors are achieved by individual calibrations *in situ*, requiring a lengthier specimen preparation. Contacting sensors require physical access to the specimen, but do not require optical accessibility. IR sensors can easily scan an observation area, giving a full picture, which cannot be obtained by contacting sensors, because a high number of sensors is not practicable.

Summarising, it may be said that IR sensors are intrinsically superior from the points of view of response time, of non-perturbation of the measured temperature, of the possibility of obtaining full images, and of the overall operational flexibility. Their signal-to-noise ratio, however, requires some specific filtering operation to reach the high resolution needed to perform good measurements of the small thermoelastic signals. Contacting sensors, in particular thermistors, have a single advantage: they can reach a high resolution without specific filtering. They are therefore suited to transient, non-repetitive signals. In the next section it will be seen that these characteristics make the two types of sensors the ideal candidates for two different types of measurements.

### 2.3.2 Experimental procedures

Mechanical tests can be performed with monotonic or cyclic (periodic or non-periodic) loading. Loads can be applied in several control modes; the load control and the strain control modes allowing a more direct interpretation of the results. In this section some observations are recalled about the influence of the load application mode on the experimental results at a given temperature. It must be remembered that the mechanical behaviour also depends on the operating temperature. This dependence can dominate for the thermally activated processes such as creep.

Referring again to the uniaxial tension or tension/compression case, the mechanical behaviour is best analysed by considering the trajectories in the stress–strain plane. In the elastic regime the same reversible linear characteristic is always followed, during both loading and unloading, irrespective of the loading mode. The response is instantaneous and linear, and the stress–strain characteristic is not modified by the loading history. The time dependence of the mechanical response, characteristic of a viscoelastic or viscous component, introduces a dependence on the type of trajectories in the stress–strain plane on the loading mode. Under stress control, creep occurs at constant stress following a stress ramp, while under strain control stress relaxation occurs at constant strain following a strain ramp. During the loading ramp itself, the trajectory deviates from proportionality, the deviation depending on the loading mode and rate. Under cyclic loading, hysteresis cycles appear. If the loading is periodic, the hysteresis cycle is stationary and its shape depends on the waveform and the cycle frequency.

The ideal plastic response is instantaneous and non-linear; the stress–strain trajectory does not depend on the loading control mode and the loading rate. The non-linearity and irreversibility of plastic strain determine the behaviour under cyclic loading. If yielding occurs during loading, plastic deformation occurs accompanied by strain hardening, so that the unloading is elastic, and further plastic strains occur only if the stress exceeds the maximum stress previously reached. Thus, if the stress is always in the same direction, plastic strain can occur continually under monotonic loading, while under cyclic periodic loading it can occur only during the first cycle. Under cyclic non-periodic loading it only occurs when the stress reaches a new maximum. If cyclic loading is performed between positive and negative stresses which both exceed the yield stress, hysteresis cycles are produced which include plastic segments. Alternating plastic strains occur which depend on the symmetry of the waveform but not on the loading frequency. In most real metals, plasticity is accompanied by a detectable amount of anelasticity. During a hold time following a plastic deformation, creep or stress relaxation can be observed. At the same time, under periodic loading which remains either tensile or compressive, the complete shakedown of plastic deformations requires more than just the first cycle. Furthermore, the stress–strain characteristic is not unique, but can be modified by alternating plastic strain; cyclic strain hardening and strain softening can occur.

The previous overview indicates that, in the linear elastic regime, measurements are not affected by the loading mode. In particular, cyclic loading can be exploited to take advantage of all the signal processing techniques developed for linear systems, as described in Chapters 3 and 6. In the anelastic case, cyclic loading can also be adopted, although linearity must be checked. Plastic deformations on the other hand, being intrinsically non-linear and irreversible, cannot exploit cyclic signal processing techniques.

The thermal response, governed by equation (2.4), is affected by thermal conduction, which is a linear, time-dependent process. Adiabatic boundary conditions are not sufficient to avoid conduction, because any stress gradient corresponds to a gradient in the effective thermoelastic (or thermoplastic) source, and therefore to a thermal gradient. Furthermore, normal specimen fixtures are often closer to isothermal than to adiabatic conditions. Thermal conduction tends to make temperatures uniform, leading to attenuated spatial temperature distributions which are smoother than stress distributions. These effects, being time dependent, are also strain (or stress) rate and frequency dependent. As it can be seen by Fourier expansion (see Chapter 4), the Fourier equation (2.4) implies a time constant $\tau$ of the order of $\tau = l^2/(\chi\pi^2)$, where $l$ is a characteristic length for heat conduction. It can typically be the distance to an isothermal boundary for nearly homogeneous stresses, or the distance over which the stress varies appreciably if stress gradients are present. For times shorter than $\tau$, therefore for frequencies above $\tau^{-1}$, nearly adiabatic conditions are achieved. The temperature amplitude distribution is then an undistorted map of the thermoelastic or thermoplastic effective source. As indicated in Chapter 1, loading frequencies above 3 Hz are required to achieve quasi-adiabatic conditions in metallic components. Care must, however, be exercised when the stress varies significantly over small distances, as in the case of high-stress concentrations.

The discussion of sections 2.3.1 and 2.3.2 finally allows the identification of the different application areas which best exploit the peculiar characteristics of contacting and non-contacting sensors. The performance of IR sensors is relatively poor on small transient signals. Filtering is also needed, which is best accomplished on cyclic signals. IR sensors are ideally suited to cyclic loading in the linear elastic regime. The other characteristics of IR sensors also lend themselves to cyclic measurements. Their fast response allows high cycling frequencies, therefore adiabatic conditions and short measurement times. They do not perturb at all the measured temperature field, and the scanning capability allows full-field images to be obtained, with all the related flexibility. Contacting techniques can give at most qualitative indications under elastic conditions, because, beside the need to mount the sensors and the impossibility of obtaining a full-field image, their slower response does not allow measurements in the adiabatic regime. Contacting techniques like thermistors may be required when small amplitude transient signals are to be quantitatively measured, because currently they are the only ones having a signal-to-noise ratio which allows good measurements without specific filtering. Measurements in the transient plastic regime exploit their sensitivity. The perturbation to the measured temperature field due to the sensors can be reduced to a negligible level, and the response times are acceptable at low to moderate strain rates. Several sensors can be mounted to obtain sampling of the temperature field at several sites. State-of-the-art IR detection coupled with the standard SPATE instrumentation is not suited to this application, due to the nature of the signal filtering. IR detection without specific filtering (AGA-Thermovision) has instead been used by Blotny and co-workers (Blotny and Kaleta 1986, Blotny *et al* 1986) in high-frequency elastic–plastic fatigue tests, in which temperature rises of the order of 100 K were observed. The rest of this chapter is devoted to the analysis of deformation phenomena, mainly in the elastic–plastic regime—the ideal application of contacting techniques.

## 2.4 MONOTONIC TESTING

### 2.4.1 Generalities

Plastic deformations are irreversible and non-linear. Therefore, as discussed in section 2.3.2, the cyclic techniques which are advantageous for linear regimes are not applicable, and monotonic tests remain the preferred ones in the analysis of plastic deformations. The thermal response in mechanical tests is governed by equation (2.4). In the absence of plastic strains $P_p = P_d = 0$, and the relationship of the thermal response to elastic strains is completely determined by the two parameters $\gamma$ and $\chi$. The same relationship between the thermoelastic part of the thermal response and the elastic part of the deformation holds in elastic–plastic tests. The Grüneisen parameter and the thermal diffusivity must therefore be measured in order to subtract the thermoelastic part of the response and single-out the thermoplastic response in elastic–plastic tests. Tests in the elastic regime allow such measurements.

The thermoplastic response is represented by $P_d$, or equivalently by the dissipation function, $f$ (see section 2.2.2), which is essentially a phenomenological parameter. A theoretical expression is not available for $f$, which depends on several parameters. The dissipation function reflects in fact the way in which the microscopic deformation mechanisms store and convert the deformation energy. The function contains information on the way the deformation micro-mechanisms are operating. This is the main reason for its measurement; it can supplement the mechanical measurements in the analysis of the plastic deformation processes.

In the first subsections of section 2.4 reference is always made to uniaxial tension tests conducted under monotonic loading. This configuration is often adopted because, beside the simplicity of the deformation state, the stress and strain state are homogeneous, or nearly homogeneous, over the useful length of the specimen. This length is typically of the order of a few centimetres. The characteristic length for heat conduction is thus of that order, and the corresponding time constant (cf. section 2.3.2) turns out to be of the order of tens of seconds. Within test durations of this order, quasi-adiabatic conditions are thus achieved, and also in longer tests the signal attenuation due to heat conduction remains limited.

Uniaxial compression tests offer the same advantage, but they are more critical, due to the possibility of buckling. However, in compression tests of cylindrical specimens, the massive compression anvils give a close approximation of well-defined isothermal boundary conditions. This configuration has been exploited by Beghi and Bottani (1980) to develop an experimental procedure to measure the thermal diffusivity and the Grüneisen parameter. A thermoelastic compression is followed by a hold time at constant strain, during which the thermoelastic temperature increase relaxes towards equilibrium. After an initial transient the relaxation becomes exponential, the time constant being controlled by the thermal diffusivity and the specimen length. The thermal diffusivity can thus be measured. Its effect can then be computed, and the Grüneisen parameter (proportional to the thermoelastic constant) is consequently accessible.

### 2.4.2 The thermoelastic–plastic limit stress

Metals such as ferritic steels present a discontinuous yielding, marked by a sharp kink in their stress–strain characteristic, and possibly by load drops under strain- or stroke-controlled tests. The transition from thermoelastic cooling to thermoplastic heating is also sharp, a marked temperature increase occurring exactly at yielding. On the other hand, metals such as austenitic steels have a face-centred cubic structure which ensures a higher number of glide planes for dislocations. Such steels exhibit a continuous yielding, the elastic to plastic transition being much more gradual. In these cases a yield stress cannot be unambiguously pin-pointed on the stress–strain curve, and conventional limits are adopted, like the 0.2% permanent strain limit. The transition from thermoelastic cooling to thermoplastic heating is also less sharp, but the slope reversal remains well marked. A minimum temperature may be seen at which thermoelastic cooling and thermoplastic heating exactly balance each other.

The minimum temperature marks the transition from a predominantly thermoelastic regime to a predominantly thermoplastic one. Bottani and Caglioti (1982a, b) have proposed the denomination 'thermoelastic–plastic limit', $\sigma_\theta$, for the stress at which the minimum temperature occurs, and its adoption as an unambiguous identification of yielding. This identification of a limit stress has the advantage of not relying on any conventionally adopted strain level. For austenitic stainless steels in the annealed state the thermoelastic–plastic limit stress turns out to be smaller than the 0.2% limit stress by an amount of the order of 10%. This difference decreases with increasing strain hardening, when the elastic–plastic transition becomes less gradual.

### 2.4.3 Deformation calorimetry: previous experimental methods

The interest in the experimental determination of the manner in which a metal undergoing plastic deformation partly stores the mechanical plastic energy and partly converts it into heat was recognised long ago. Many measurements have been performed by annealing calorimetry techniques. The basis of such techniques is simple; the energy which is stored during a deformation is 'frozen' in the solid and can be liberated by annealing, which promotes recrystallisation. The energy which is liberated can be detected as a modification of the apparent specific heat. Several sophisticated procedures have been devised. The comprehensive review by Bever *et al* (1973) presents many details, later works always relying on the same principles. These methods can reach very good sensitivities. Their main limitation is in the fact that they measure the integral amount of energy $Q_i$ which has been stored during a deformation phase, but cannot give any indication on how this storage evolved during the deformation.

A different approach to the problem is that of deformation calorimetry, in which the amount of heat dissipated during the deformation is measured during the deformation itself or in an immediately subsequent stage, without modifying the state of the specimen. Deformation calorimetry techniques encounter greater difficulties in achieving good resolutions, but have a greater potential because they allow consecutive measurements to be performed. The effect of incremental deformations is evaluated and procedures can be devised to measure the incremental amounts of energy which are dissipated at any stage during the deformation, and also possibly after deformation is stopped.

In their pioneering work, Farren and Taylor (1925) adopted uniaxial tension specimens and fast deformations, in order to measure the adiabatic temperature rise immediately after the end of deformation. They measured the specimen temperature by a thermocouple, and carefully selected the type of mounting in order to obtain a fast response. Taylor and Quinney (1934) performed similar measurements, in which they adopted the same temperature measurement technique, but in which they used torsion specimens and compression specimens. The amount of plastic deformation occurring before rupture, and consequently the amount of converted and stored energy, is thus significantly larger than in the tensile case.

Williams (1960) developed a calorimeter in which a sample undergoes a compressive impact deformation, its temperature being measured by a thermocouple. Heat leakage from the specimen is minimised by working in a vacuum chamber and by suspending the specimen from the thermocouple itself. A careful signal suppression allows for good sensitivity. It was thus shown that a detectable amount of energy is released during a time period following the deformation. The apparatus was exploited (Williams 1961) to study the influence of metallurgical variables on the amount of stored energy. In another experimental arrangement (Williams 1964) tensile specimens were used; heat leakages were minimised by adopting nylon pull rods. Williams (1963) also developed a different type of calorimeter, in which the specimen was wetted by a liquid film. The amount of evolved heat was measured by the volume of the vapour which was generated under conditions near to equilibrium. Wolfenden and Appleton (1967) developed a calorimeter which exploited the same principle; Wolfenden (1968) used this calorimeter at temperatures down to 78 K.

Ronnpagel (1979) developed a calorimetry method in which a tensile specimen was deformed under non-adiabatic conditions, its temperature being sensed by a thermistor. Heat leakage was de-convoluted by a response function, which was determined by measuring the temperature signal during elastic unloadings and reloadings. The response function was repeatedly determined during the tests. Ronnpagel and Schwink (1978) exploited the method to perform measurements on copper single-crystals. Cernocky and Krempl (1981) and Krempl (1985) measured the specimen temperature using several thermocouples during tension–compression testing of steel specimens in the elastic–plastic regime. The cycling frequency (typically 1 Hz) guaranteed nearly adiabatic conditions for the first cycles; a complete analysis was performed only for this initial segment of the tests. Chrysochoos (1985) developed a calorimeter in which the specimen temperature was not directly measured, but heat leakage from the specimen was measured by the temperature drop across a controlled boundary; tests were performed on commercial alloys.

### 2.4.4 Deformation calorimetry: the experimental method developed at CESNEF†

An experimental method of deformation calorimetry has been developed at CESNEF. The method exploits the performance advantages of thermistors in the measurement of transient temperature signals, and is aimed at the determination of the instantaneous values of the thermal powers per unit volume; 'differential' or 'incremental' calorimetry can therefore be performed.

The method is based on equation (2.4), which can be rewritten as

$$\frac{\partial\theta}{\partial t} - \chi\nabla^2\theta = -\gamma T_0\frac{\partial(\delta\varepsilon)}{\partial t} + \frac{P_d}{\rho C_v} = \frac{P_{th}}{\rho C_v} \tag{2.5}$$

where $P_{th}$ is the sum of the effective heat sources due to the deformation processes. The temperature-field maps $P_{th}$, apart from the heat capacity per unit volume, $\rho C_v$, which is assumed to remain constant during the test. The mapping is distorted by thermal conduction. Only in very peculiar cases, as in the experiments performed by Williams (1960, 1964), can conduction be neglected; in most cases it must be de-convoluted. Ronnpagel (1979) developed an integral de-convolution procedure, relying on a global transfer function between the heat source in the specimen and the temperature measured at a given site. The method developed at CESNEF relies on a differential de-convolution procedure, in which the Laplacian of equation (2.5) is evaluated.

If the temperature can be measured with sufficient resolution, accuracy and spatial resolution, the derivatives on the left-hand side of equation (2.5) can be measured, and $P_{th}$ is obtained via the two constants $\chi$ and $\rho C_v$. This measurement can be considered for two-dimensional surface-temperature fields. However, on transient signals only the contacting sensors can reach the resolution needed for this application, and the sampling of a two-dimensional field requires a number of sensors which is too high to be considered practicable. Only the sampling of one-dimensional fields can be considered in practice. Nevertheless, the temperature field in a tensile specimen undergoing a tensile test depends, to a good approximation, on the longitudinal co-ordinate only, because of the space distribution of the

† Centro Studi Nucleari Enrico Fermi.

effective source and of the thermal boundary conditions. Firstly, the deformation field is typically homogeneous, except when necking; secondly, a typical tensile specimen has relatively massive heads in good thermal contact with the fixtures. Isothermal conditions are therefore approximately maintained within the heads. The central part of the specimen exchanges heat with the ambient air and with the specimen heads. The values of the heat exchange coefficient with the ambient air at rest (e.g. McKelvie 1987) and of the thermal conductivities of metals are such that heat exchange with the ambient air is negligible with typical specimen geometries. Radial gradients are thus negligible. The temperature field is therefore essentially one-dimensional, and lends itself to spatial sampling by thermistors.

Under uniaxial tension conditions, the plastic mechanical power is $P_p = \sigma(\varepsilon - \sigma/E)$, while the volume variation $\delta\varepsilon$ is given by $\sigma(1 - 2v)/E$ (section 2.2.2). The conventional stress and strain measurements thus give immediate access to $P_p$ and to the first term on the right-hand side of equation (2.5). If $P_{th}$ can be measured by the derivatives of the temperature field, $P_d$ is immediately extracted, and can be compared to $P_p$. The temperature field $\theta(x, t)$ is therefore measured, during a tensile test, by several thermistors located at longitudinal positions $x_i$, equi-spaced along a tensile specimen (see figure 2.2(a)). The $\theta(x_i, t)$ values are thus obtained. Although the derivatives $\partial\theta/\partial t$ and $\partial^2\theta/\partial x^2$ can be evaluated by incremental ratios, the estimates obtained in this way are affected by unacceptable amplifications of the noise component of the signals; smooth estimators of the derivatives must be found. Suitable estimators can be obtained by considering an estimated temperature field which obeys the appropriate conditions, and by fitting it to the measured temperatures by a least-squares procedure. The estimated temperature field must not rely on any assumption about the longitudinal boundary conditions, because boundaries cannot be identified at which precise boundary conditions can be assigned, and because the temperature measured at any arbitrary boundary is affected by the same level of uncertainty and noise as the temperatures measured at any other site. To date, estimation procedures of this kind have been found only under the hypothesis that $P_{th}$ is longitudinally homogeneous, i.e. that the deformation is homogeneous. In this case an estimated temperature field can be found by an instantaneous approximation by spline functions, or by a more general two-dimensional fitting procedure, in which the estimated field is forced to satisfy the discretised form of equation (2.5). This condition is imposed locally, and does not rely on any assumption about either the boundary conditions or the functional form of the temperature field. An example of the time evolution of the spatially homogeneous effective source, $P_{th}$, obtained by the above procedure, is presented in figure 2.2(c).

Macroscopically homogeneous deformation occurs in ductile metals before the initiation of necking phenomena. In particular, for face-centred cubic metals which exhibit a continuous yielding, deformation is also

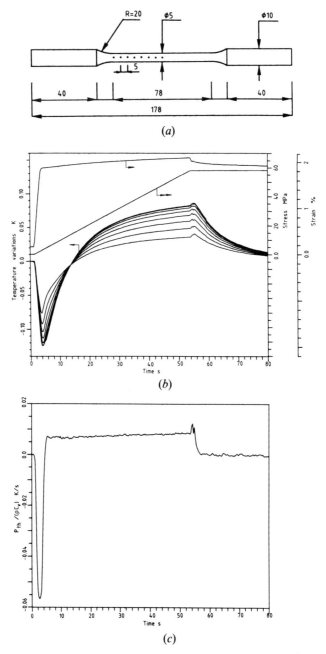

**Figure 2.2** Strain-controlled tensile test of an aluminium specimen. (*a*) Specimen dimensions and positions of eight thermistors; (*b*) stress, strain and temperature variations measured during the test; and (*c*) thermal source obtained from the temperatures of figure 2.2(*b*).

sufficiently homogeneous in the yielding phase. Metals exhibiting a discontinuous yielding undergo non-homogeneous deformation in the yielding phase. Accordingly, to date the method has been exploited with metals like pure aluminium and pure copper. Cylindrical specimens are used, the geometry of which is shown in figure 2.2($a$). Tests in the elastic regime, in which $P_d = 0$, allow the measurement of the constant $\gamma(1 - 2\nu)$, and thus the determination and subtraction of $P_{te}$ in the elastic–plastic tests. The thermal diffusivity $\chi$ is consistently measured by a similar procedure applied to a relaxation phase at constant strain in which $P_{th}$ is zero. The temperature is initially not at equilibrium, due to a previous deformation phase, but it approaches equilibrium, the relaxation being governed by $\chi$.

The temperatures $\theta(x_i, t)$ are measured by miniaturised ($\simeq 0.3$ mm) bead thermistors glued on one-half of the specimen. The rest of the specimen is available for the strain sensor. Eight thermistors are typically used. This number turns out to be a good compromise between resolution and experimental simplicity. As the space derivatives of the temperature field are to be estimated, reliable measurements of small differences among temperatures detected by different sensors are crucial. To maximise and to assess this reliability, the thermistors are individually calibrated *in situ* by comparison with a platinum resistance thermometer, and the calibration is tested by measuring the slow drifts of the ambient temperature of the laboratory. The specimen is enclosed in a box which is shielded from air currents, and the metallic specimen provides a good thermal equilibrium to the sensors. The different channels measure this type of temperature variation, which is identical for all the sensors, with differences which are typically around 1% of the instantaneous variation. Considering then the absolute accuracy of the measurement, its assessment is limited by the resolution attainable by the platinum resistance thermometer, and turns out to be better than 3% of the instantaneous variation. The temperature is measured during conventional tensile tests, conducted on a standard material testing machine (Instron 1121), equipped with standard load and strain measurement systems. Tests are conducted under either stress or strain control, the typical load history being formed by a series of trapezoidal cycles.

### 2.4.5 Deformation calorimetry: results

Bever *et al* (1973) conducted an extensive survey of results obtained by annealing calorimetry and by deformation calorimetry experiments, and discussed the possibility of comparing the results obtained by the two methods. The various experimental results are relatively scattered, and the significance of comparisons among different experiments is limited by the appreciable influence of the purity of the material and of the type of crystal (single crystal or polycrystal). In most cases energy data are presented, which refer to relevant amounts of deformation (typically one or several per cent).

The instantaneous values of power, which can be obtained by the present experimental method, are seldom obtained by other methods. The average values of the dissipation function, $f$, can be compared; the present method leads to values which are typically around 70–80% for 99.5% pure copper and aluminium specimens. These values are comparable to those obtained by Chrysochoos (1985) for industrial alloys. High-purity materials and single crystals give values of $f$ in excess of 90% (Bever *et al* 1973).

A typical result obtained at CESNEF on a 99.5% pure aluminium specimen tested under strain control is presented in figure 2.2(*b*). The transition from thermoelastic cooling to thermoplastic heating is evident. Strain hardening is very limited; the flow stress remains nearly constant. Consequently, under strain control, the elastic strain rate and the thermo-elastic effective source, $P_{te}$, are almost zero, the total thermal source, $P_{th}$, nearly coincides with the dissipated power, $P_d$, the plastic mechanical power $P_p = \sigma \dot{\varepsilon}_p$ remains nearly constant, and $P_d$ also turns out to remain nearly constant. The temperature rise and the corresponding thermal source peak at the end of the ramp are due to the non-ideal operation of the control unit, which causes a partial elastic unloading at strain arrest.

In figure 2.3(*a*) results are presented for a 99.5% pure copper specimen, previously deformed to about 4%, and loaded under stress control. Copper exhibits an easy glide region, followed by a more marked strain hardening. Consequently, under stress control, yielding involves a marked increase in the strain rate, followed by a partial decrease. Accordingly, $P_p$ has a peak, while $P_d$ sometimes has (figure 2.4(*b*)) and sometimes does not have (figure 2.3(*a*)) a similar peak. The different behaviour seems to be correlated with different initial conditions of the specimens. The other oscillations in the results are essentially artefacts, related to numerical differentiation of sampled data. The trend is, however, well determined. Figure 2.4 presents the results from a whole load history, formed by several trapezoidal stress cycles at increasing stresses, imposed on a previously annealed copper specimen. At permanent strains of about 4% and 10%, several consecutive cycles are performed with the same maximum stress. Elastic ramps are thus obtained, which are exploited to measure the Grüneisen parameter. Anelastic relaxations at constant maximum stress, which are a manifestation of a viscoelastic component, are noticeable in the last cycles. The dissipation function (figure 2.4(*c*)) is low at very low strains, when the material hardens considerably by significant modifications of its dislocation content. At higher strains, the modifications of the microstructure become more gradual, and the dissipation function grows to values close to 80% which do not exhibit significant variations over a wide strain interval.

### 2.4.6 Applications to the study of deformation mechanisms

The energy storage and conversion can be analysed within the framework of continuum mechanics (Krempl 1985, Chrysochoos 1985), as well as in the

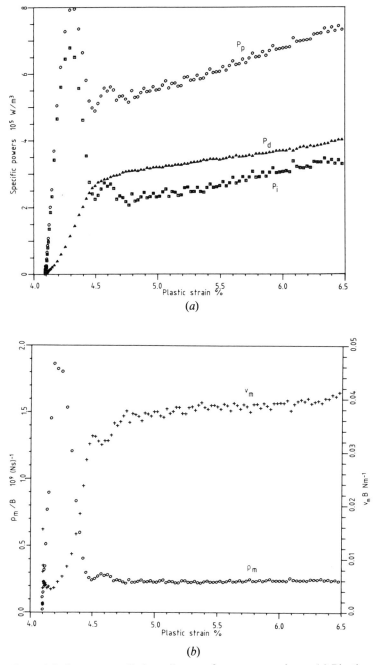

**Figure 2.3** Stress-controlled tensile test of a copper specimen. (*a*) Plastic
($P_{\mathrm{p}}$), dissipated ($P_{\mathrm{d}}$) and stored ($P_{\mathrm{i}}$) powers at and after yielding; and (*b*)
dislocation density ($\rho_{\mathrm{m}}$) and velocity ($v_{\mathrm{m}}$) obtained from figure 2.3(*a*).

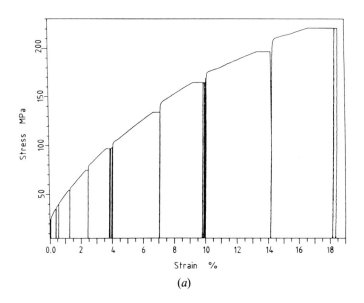

(*a*)

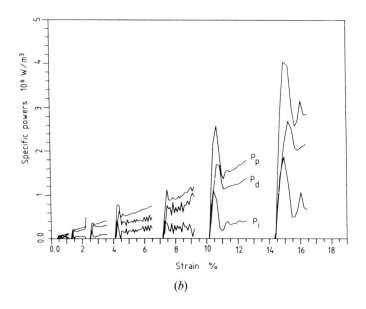

(*b*)

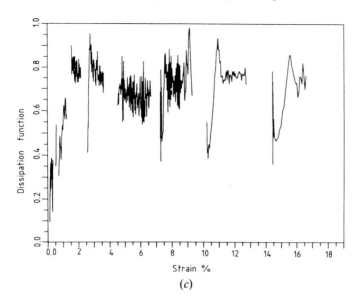

(c)

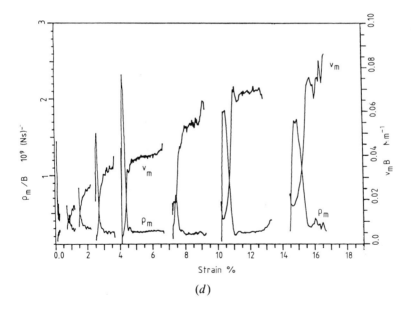

(d)

**Figure 2.4** Stress-controlled tensile test of a copper specimen subjected to repeated loadings (6.5 MPa s$^{-1}$). (a) Stress–strain diagram for the whole load history; (b) plastic ($P_p$), dissipated ($P_d$) and stored powers ($P_i$); (c) dissipation function ($f$); and (d) dislocation density and velocity.

framework of microscopic theories of deformation. Focusing on the micro-mechanisms of plastic deformation (i.e. dislocation motion and creation), the intrinsically macroscopic energy data can be exploited to study the evolution of parameters which represent the average evolution of the dislocation network.

The Orowan's relation (see e.g. Cottrell 1953) links the macroscopic plastic strain rate to the average density of mobile dislocations $\rho_m$ (dislocation length per unit volume) and to their mean velocity $v_m$:

$$\dot{\varepsilon}_p = mb\rho_m v_m . \tag{2.6}$$

Here $b$ is the modulus of the Burgers vector of dislocations and $m$ is Schmid's geometric orientation factor. A model is needed to link the macroscopic energy data to similar averages of microscopic parameters. A simple model can be considered, in which dissipation is assumed to occur during disloca-tion motion only. If furthermore the drag force on dislocations (per unit length of dislocation) is assumed to be proportional to the dislocation velocity via a microscopic viscous drag coefficient, $B$, the dissipated power can be expressed as

$$P_d = (Bv_m)v_m\rho_m . \tag{2.7}$$

Equations (2.6) and (2.7) can be solved with respect to $v_m$ and $\rho_m$, obtaining

$$v_m = \frac{mbP_d}{B\dot{\varepsilon}_p} \qquad \rho_m = \frac{B(\dot{\varepsilon}_p)^2}{(mb)^2 P_d} . \tag{2.8}$$

It is thus possible to derive the average values of the dislocation density and velocity from the macroscopic quantities $\varepsilon_p$ and $P_d$, which are experimentally accessible. The coefficient $B$ can be considered a phenomenological coeffi-cient, the direct measurement of which is difficult. However, its value only represents a scale factor, and the behaviour of $\rho_m$ and $v_m$ is completely determined, apart from this scale factor.

The results obtained from the previously mentioned data are presented in figures 2.3(b) and 2.4(d) respectively. The yielding of copper turns out to be associated with a sharp rise of the mobile dislocation density, the mean velocity remaining low. The end of the easy glide phase is associated with a marked decrease of the mobile dislocation density (dislocations are blocked), while the mean velocity increases. The density remains approximately constant in the strain hardening phase, while the velocity slowly increases. To date, these observations are essentially qualitative since in the yielding phase, in which a large number of dislocations are mobilised or created, the assumption of the model (dissipation only due to dislocation motion) is rather questionable. The measured dissipated power does, however, repre-sent experimental evidence which can be compared with any model.

### 2.4.7 Non-homogeneous deformation fields

In more complex geometries, the stress and strain fields cannot be measured. Finite element computations are commonly adopted to determine them. Stress and strain fields in the elastic–plastic regime can be computed by non-linear finite element codes, which rely on accepted models of plastic flow, concerning the yield surface and the flow rule. These computations are usually validated by a comparison with measurable quantities, such as the relative displacements of given points.

Temperature measurements offer an additional opportunity to compare computations and measurements. Elastic–plastic finite element computations also lead to the local values of the volume variation $\delta\varepsilon$ and of the mechanical plastic power $P_p$. The space distributions of the thermoelastic and thermoplastic effective sources $P_{te}$ and $P_d$ can thus be computed, if the Grüneisen parameter and the dissipation function $f$ are known. The Fourier equation (2.5) can then be integrated, again by a FE technique, obtaining a computed temperature field, which can be compared with measured temperature variations. This comparison represents a global test of the ensemble of the hypotheses adopted in the computations. The Grüneisen parameter is a measurable constant (see section 2.4.4), while the behaviour of the dissipation function is not known. It can depend, among other factors, on the stress or the strain, the strain rate, the triaxiality of the stress, or the temperature. In order to demonstrate the outlined procedure, and in the absence of detailed information about $f$, an experiment and a computation are performed, in which the simplest assumption is adopted; $f$ is assumed to be a constant.

A Compact Tension specimen was used (figure 2.5($a$)) with a notch stress concentration. A common low-carbon ferritic steel was chosen (Fe37, UNI7070-72), the stress–strain characteristic and Grüneisen parameter of which were measured on tensile specimens machined from the same batch of material. The notch radius was 1.0 mm; a sharper notch or a crack was avoided, due to possible difficulties in the FE computation. A thin (6 mm) specimen was used, to approximate plane stress conditions, in which the stress and temperature fields at the surface do not differ appreciably from those in the bulk of the specimen. A load ramp was applied, which produced yielding and plastic flow in the notch region. The ramp lasted 30 seconds, and it reached a maximum load at which the load–COD (crack opening displacement) characteristic deviated significantly from linearity (figure 2.5($b$)) and the predicted plastic zone had a width of several millimetres. In the notch region, the temperature was measured by ten thermistors, glued to the specimen and calibrated *in situ* by the procedure described in section 2.4.4.

The FE computation was performed by a commercial code, which models the plastic flow by the von Mises yield criterion and the Prandtl–Reuss flow rule. The stress–strain characteristic was modelled by two linear segments.

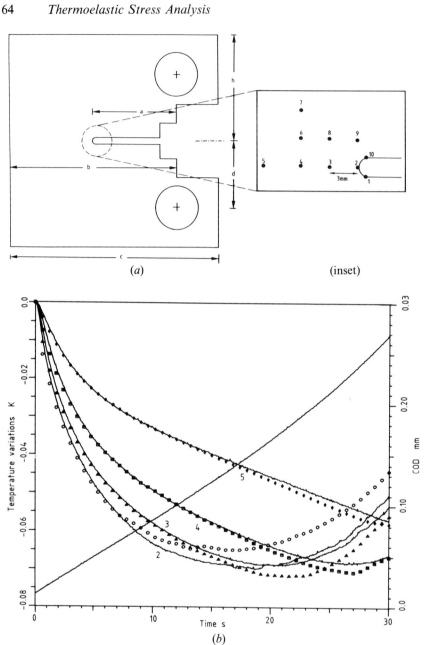

(a)                                    (inset)

(b)

**Figure 2.5** Load-controlled test of a ferritic steel specimen (250 N s⁻¹).
(a) Specimen dimensions ($a = 25$ mm, $b = 50$ mm, $c = 62.5$ mm, $d =$
18.75 mm, $h = 30$ mm, thickness = 6 mm, radius of notch = 1 mm) and
(see inset) thermistor positions; and (b) crack opening displacement
(COD) during the load ramp and computed and measured temperatures at
locations 2, 3, 4 and 5.

The computation was performed with small load increments, to obtain a detailed representation of the time evolution of the $\delta\varepsilon$ and $P_p$ fields. An interface programme was developed to compute the thermoelastic and thermoplastic effective sources as functions of space and time. A commercial FE code was finally used to integrate the Fourier equation with the computed source, and obtain a predicted temperature field. Adiabatic boundary conditions were assumed, since heat conduction between stressed and unstressed regions is far greater than heat exchanges with the calm ambient air.

The computed load–COD characteristic turns out to be in good agreement with the measured characteristic. Preliminary tests at moderate loads, in which the deformation is exclusively elastic, show a very good agreement between thermal computations and measurements. Figure 2.5(*b*) presents the results from the load ramp. The agreement between the computed and measured temperatures is excellent in the elastic phase, while deviations are noticeable in the surroundings of the notch (thermistors 2 and 3) when plastic flow occurs. In particular, at thermistor 2 the thermoplastic heating starts later than predicted, while at thermistor 3 it starts earlier than predicted by the computation. At higher distances from the notch (thermistors 4 and 5) the agreement remains good. A constant value of 75% is adopted for *f*. This value is in agreement with those found in the calorimetric tests (although for different materials), and gives, among a small set of tested values, the best overall agreement between computation and measurement.

The incomplete agreement between computation and measurement in the notch zone can be ascribed to several causes. Firstly, two linear segments are a poor approximation of the stress–strain characteristic in the post-yield phase, in which deformation occurs at constant stress before strain hardening sets in. Secondly, the plane stress conditions are only approximately achieved, and triaxiality effects may be present. This observation is supported by the fact that a test performed on a thinner (4 mm) specimen gives a better agreement. Thirdly, the representation of *f* by a constant value is very crude, as *f* has most probably a more complex behaviour. The ensemble of the results indicates that, despite the mentioned uncertainties and approximations, the adopted models of the plastic deformation and the thermal effects lead to a computed temperature field which is in reasonable agreement with the measured field, the agreement being excellent in the elastic case. While this agreement does not offer a substantial new validation to the model of plasticity, it does indicate that approximate predictions of temperature fields can be obtained with very simple models of the thermoplastic effect.

## 2.5 CYCLIC TESTING

### 2.5.1 Generalities

The Fourier equation (2.5) is linear, and the thermoelastic effective source is a linear function of the local strain. The relationship between the temperature and the elastic strains is thus linear, as well as the stress–strain relationship. This is the basis of thermoelastic stress analysis techniques. The relationship between the temperature and the plastic strains is non-linear, as is the stress–strain relationship. In section 2.4 situations have been considered in which macroscopic yielding occurs and the non-linear plastic effects are dominant. In the case of a cracked specimen under cyclic loading a different situation occurs. Most of the specimen behaves elastically, while non-linear deformation and damage processes are active in the crack tip zone. The macroscopic (both mechanical and thermal) behaviour is then essentially linear, with small deviations from linearity. In order to observe the non-linear effects, the main objective of a measurement is the discrimination of a small non-linear component from a predominant linear response.

To this purpose, the signal analysis procedures described in section 6.2 are best suited. Their applicability is not ensured because, if a damage process is active, the state of the specimen is continuously modified, and the measured signals are not stationary. However, if the damage process is slow enough, the signal characteristics evolve slowly, and the spectral analysis techniques remain applicable. In practice the Fourier transform remains meaningful if, during the time window required to perform a reliable transform, the modifications of the spectrum of a signal remain of the order of the uncertainty in the determination of the spectrum itself. Under uniform cyclic loading this condition is typically met for crack propagation rates up to around 0.001 mm per cycle. When the spectral analysis is applicable, uniform sinusoidal loading is best suited to discriminate the linear and non-linear components of the responses, because the linear component falls at the excitation frequency, while the non-linear components give contributions at other frequencies. The separation is, however, not complete because a non-linear periodic response can also contain a component at the excitation frequency.

### 2.5.2 Tests on cracked specimens

Tests were performed on an austenitic stainless steel (AISI316) Compact Tension specimen (figure 2.6(a)). The specimen was fatigue pre-cracked and eight thermistors were then glued onto its surface around the crack zone, according to the procedures described in section 2.4.4. Subsequent load cycles were monitored. A wide specimen was chosen, so that a crack growth of a few millimetres still represented a small fraction of the ligament. The growth thus occurs without substantial modifications of the stress field in the zone surrounding the crack tip, and with a slowly varying rate. The specimen was thin (8 mm), and further thinned to 5 mm in the crack zone, to achieve plane stress conditions, in which the surface temperature is representative of the whole thickness. Loading was performed under load control, at a frequency of 0.2 Hz. A low frequency was selected to avoid distortions of the temperature signals due to the frequency response function of the sensors. Heat conduction therefore unavoidably occurs, and the temperature gives a distorted map of the effective thermal source. Figure 2.6 reports a typical measurement; figure 2.6(*a*) presents the approximate position of the crack at the stage in which the data of figure 2.6(*b*) were recorded.

A sinusoidal load was imposed; the load–COD characteristic is linear, with hysteresis cycles of very small area. The temperature responses are sinusoidal, with different amplitudes and phases, and are affected by small distortions of the purely sinusoidal waveform. The phase differences are produced by thermal conduction, the strain and the thermoelastic effective source being non-homogeneous. The amplitudes are related, via the distortion introduced by thermal conduction, to the thermoelastic effective source distribution. The distortions are best analysed by the auto-power spectra of the signals (see section 6.2). The load has a good spectral purity, while the small hysteresis cycles manifest themselves in the power spectrum of the COD by small peaks at multiples of the loading frequency. The temperature power spectrum (figure 2.7) is also characterised by peaks at the exact multiples of the loading frequency. These peaks are significantly larger than those in the spectrum of the COD. The presence of discrete peaks at the harmonics of the excitation frequency, without noticeable contributions at other frequencies, indicates that the signal can be expanded in discrete Fourier series. This means that the signals are nearly perfectly stationary, and that they can be characterised by the weights of the components at the discrete harmonics.

Over 10 000 cycles have been monitored, corresponding to a crack growth of 5 mm. The power spectra turn out to be consistently measurable, although for the smallest peaks, which correspond to very small signal amplitudes, the signal-to-noise ratio is poor and the reproducibility is limited. The weights of the higher harmonics decrease rapidly at increasing distances from the crack tip, indicating a strong localisation of the non-linear phenomena in the process zone at the tip. In particular, the harmonic content at a given site continuously increases when the crack approaches the site, while it decreases at the sites behind the crack tip. The interpretation of the results is, however, not obvious since non-linearities can also be produced by geometrical effects. The lips of a crack are a unilateral constraint, which transmits compressive stresses only. Therefore, if crack closure phenomena occur, the relationship between the applied load and the local stress, and consequently the strain and temperature, is non-linear, even when the material behaves elastically.

In conclusion, the small non-linear component of the signals can be detected and can be characterised in a reproducible and consistent way. It carries information about the damage processes at a crack tip, but still needs a clear interpretation.

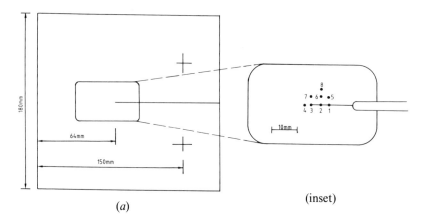

(a)                                    (inset)

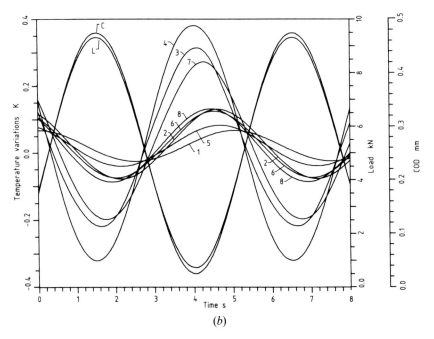

(b)

**Figure 2.6** Load-controlled test at 0.2 Hz of a fatigue-cracked austenitic stainless steel specimen. (a) Specimen dimensions and position of thermistors around the notch (see inset); and (b) load, crack opening displacement (COD) and temperature variations.

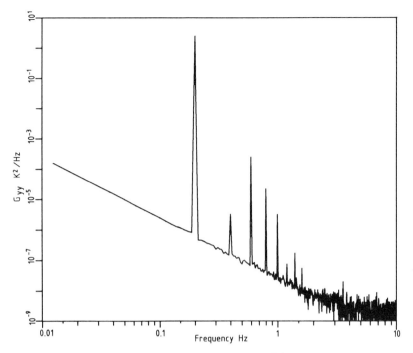

**Figure 2.7** Sample auto-power spectrum of the temperature.

## 2.6 CONCLUSIONS

The different deformation regimes of metals have been reviewed, both from a mechanical and a thermodynamic point of view. The thermal effects of deformation turn out to offer a useful tool for the analysis of deformation mechanisms. Contacting and non-contacting techniques for temperature measurements have been considered; while the non-contacting measurements are best suited in the elastic regime, thermistors are needed to make quantitative measurements of the transient signals produced by the thermo-plastic effect.

Three main applications have been considered which take advantage of the performance of thermistors and are affected by their limitations. Simple geometries like tensile specimens allow calorimetric measurement, by which the thermoelastic and thermoplastic effects can be characterised, and which can be exploited to study the deformation micro-mechanisms. Geometries in which complex stress patterns occur can be used to compare the predictions of FE computations with measurements. The presence of a crack introduces a small localised non-linearity in a globally elastic behaviour; this non-linearity has been detected and characterised.

## ACKNOWLEDGMENT

The author is grateful to Elsevier Science Publishers, Amsterdam, for permission to reprint material published in *Materials Letters* **6** 133–7 (1988).

# TSA Signal Processing and Data Collection

## B R Boyce

### 3.1 INTRODUCTION

To apply Thermoelastic Stress Analysis (TSA) technology properly, an understanding of the necessary signal and data processing is important. In Chapter 1 the fundamentals of TSA were explained. The thermoelastic theory and the physics of IR detection information from Chapter 1 is augmented in this chapter by an explanation of how IR detector signals can be processed for stress analysis applications.

Ometron Ltd makes the only TSA instrument currently marketed: the SPATE 9000. In Chapter 1 a brief historical review tells how the SPATE 9000 came to be. A few more details of that evolution are presented here for background.

The SPATE 9000 was preceded by Ometron's original TSA product: the SPATE 8000. The 8000 differs from the 9000 primarily in two ways:

(i) The 9000 is capable of collecting two channels of information about the TSA signal: the amplitude and phase. The 8000 could only collect one channel: the amplitude.

(ii) The 9000 is a modifiable instrument system base on a Hewlett Packard workstation computer, while the 8000 was a turnkey system based on a dedicated Intel microprocessor. The software system for the 9000 is extensive and can be added-to by the user.

The evolution of the 8000 to the 9000 was the result of a co-operative effort by Ometron, the University of Wisconsin, and the John Deere Horicon Works (located in Wisconsin, USA).

The SPATE 9000 can take many forms and be used in a broad range of applications. This chapter will cover only the most common use of thermoelastic technology. Other chapters in this book cover the application of TSA to a number of engineering fields. In reading those chapters the reader will learn about special implementations of TSA technology.

## 3.2 SIGNAL PROCESSING AND ELECTRONICS

As mentioned above, the current state of the art in thermoelastic stress analysis is represented by the Ometron SPATE 9000 instrument. Figure 3.1 shows a block diagram of Ometron's SPATE 9000 system. The 9000 uses a single IR detector and a set of scanning mirrors to make measurements of thermal oscillations on the surface of a target specimen. Its primary use is for measuring the stress over the surface of a specimen loaded cyclically at a single frequency of constant amplitude.

The optical head of the SPATE 9000 system contains an IR detector and lenses, a beam splitter, visible light optics, and scanning mirrors. Light enters the head via the scanning mirror system aimed by the computer. The light is then split into IR and visible light channels. The IR optics channel focuses onto a cryogenically cooled (77 K) detector. The visible light channel delivers an image through an eye-piece to aid in aiming the IR optics.

Cyclic loading of a target specimen causes a cyclic IR detector output. The amplitude of this AC signal is converted to a DC level by a lock-in amplifier. (The lock-in amplifier is also commonly referred to as a correlator or phase-sensitive detector, PSD.) The computer system uses an analogue-to-digital converter (ADC) to sample the lock-in output which is then taken as the stress amplitude for the targeted point on the specimen.

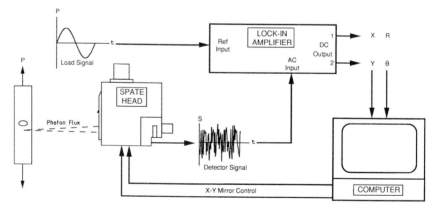

**Figure 3.1** Block diagram of the SPATE 9000 system.

In the most common application of TSA techniques the computer is programmed to move the scan mirrors from place to place in a raster pattern, making many measurements. These measurements are then presented as a colour-enhanced image of the specimen surface stress. Another common mode of operation for the instrument is the measurement of stress along a line or group of line segments. A less common application of the instrument is to track the time-dependent change of the cyclic stress amplitude at a point or group of points.

The following sections of this chapter will discuss and describe the collection and analysis of thermoelastic data.

### 3.2.1 The detector signal

Consider the output from the detector of a TSA system. For a fairly high cyclic stress in a steel specimen it is typical to have a detector output signal of the order of 0.020 V peak-to-peak. Unfortunately, this signal will be buried in a 0.800 V peak-to-peak noise signal of very broad bandwidth. This poor signal-to-noise ratio is primarily a consequence of the nature of the detector and is inherent in the detectors available today.

It is very important in TSA work that measurements of the detector output signal be made quickly. This is because many measurements need to be made so as to develop an image or stress pattern. The magnitude of this consideration can be illustrated with a few calculations. A reasonable image will probably require at least 100 by 100 measurements or pixels (picture elements). In the Ometron system the image may be as large as 256 by 256 measurements. This means that between 10 000 and 65 000 measurements are made to develop a surface stress image. If the specimen is loaded at 10 Hz and one load period is required for each amplitude measurement, then to complete the scanning process would take between 1000 and 6500 seconds, or from 17 to 110 minutes.

Fortunately, a lock-in amplifier is exceptionally well suited to the task of finding a signal of a known frequency buried in random noise. The lock-in is adjustable so that there is a direct trade-off between time to make a measurement, and noise. The slower the measurement, the more precise it will be.

### 3.2.2 The lock-in amplifier

The lock-in and its capabilities are crucial to the SPATE 9000 instrument. This section explains some of the fundamental functions and characteristics of the lock-in. More detailed information than it is possible to supply in this chapter may be obtained from the manufacturers of the instrument, EG&G Brookdeal Electronics, Princeton, NJ.

The lock-in amplifier in its simplest form (figure 3.2) is a signal mixer and a low-pass filter connected in series. The signal mixer combines the detector and a reference signal by forming the product of the two signals. When two signals are combined in this way the resultant signal contains the sum and difference of the input frequencies. For TSA this is the sum and difference of the detector and reference frequencies. Notice that the portion of the detector signal of interest and the reference signal have exactly the same frequency. Therefore the difference component of the mixer output is represented as a DC level. Moreover this DC level is proportional to the amplitudes of the detector and reference signals. The actual lock-in instrument has circuits that normalise the reference signal so that the DC level is proportional to only the amplitude of the detector signal (at the frequency of the reference signal). An AC component at the sum frequency is also present in the mixer output, but this will be filtered out in the next stage of the lock-in amplifier.

Any noise present in the detector signal that is not at the reference signal frequency is also multiplied by the reference and presented at the output of the mixer. However, by passing the output of the mixer through a low-pass filter most of the noise of the detector signal is effectively integrated out. The lock-in output has a very low equivalent noise bandwidth by comparison to its input. (For now we will use equivalent noise bandwidth as a measure of the noise, and to help compare noise elimination techniques. Equivalent noise bandwidth may be thought of as the 20 dB down point or the 10%

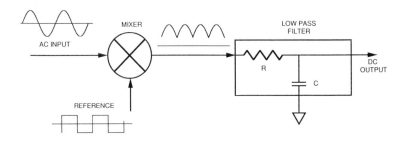

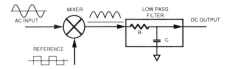

**Figure 3.2** A simple lock-in amplifier.

signal point of a low-pass filter.) The output noise bandwidth of the lock-in is a function of the low-pass filter time constant, $T$. For a typical two-pole low-pass filter with a roll-off of 12 dB/octave, the output noise bandwidth, $f_n$, is

$$f_n = 0.125/T. \tag{3.1}$$

Here a time constant of 0.1 s translates to a 1.25 Hz bandwidth. For a 0.1 s time constant setting on the lock-in, only the detector noise signals within 1.25 Hz of the reference will appear on the output of the lock-in.

An actual lock-in amplifier often has two output channels, $R$ and $\theta$, that represent the amplitude of the input and its phase relationship to the reference signal, respectively. More commonly used is the alternative presentation of $X$ and $Y$, the in-phase and quadrature components of the input signal. This representation generally works better for TSA work. This is because the lock-in can be phase-adjusted so that the $Y$-component is a minimum, making the magnitude of the $X$-component equal the amplitude. The sign of the $X$-component can be interpreted into stress sign information (tension or compression). Another reason for setting up the lock-in to produce $X$- and $Y$-components rather than amplitude and phase is that as the amplitude gets small the phase signal gets very noisy. In fact the circuitry of the lock-in is such that $X$ and $Y$ are the more fundamental measurements.

It can be very important in acquiring TSA data that both the $X$ and $Y$ signals are sampled and recorded. This is because one of the main criteria of making good TSA measurements is maintaining adiabatic conditions (or at least limiting the effects of non-adiabatic conditions). If conduction of heat through high-stress gradients (non-adiabatic conditions) are significant in a particular TSA scan, then there will be a tell-tale phase shift. This is because conduction acts like a filter, reducing the amplitude and shifting the phase of the thermal signal. Paint thickness is another factor that may cause troublesome phase shifts and signal attenuation, as described in Chapter 4. Modal behaviour may also produce phase shifts, as shown in Chapter 6.

### 3.2.3 Sampling the lock-in amplifier

Using the lock-in amplifier with a scanning mirror system puts some stiff constraints on making thermoelastic stress measurements. As mentioned earlier, a large number of data points are needed to construct a stress map of the specimen surface, so we are interested in making measurements as fast as we can. Because of this we need to be concerned about how long it takes for the scan mirrors to move and for the lock-in output to come to a new reading. These constraints mean that the sum component of the mixer output that is at twice the frequency of the reference signal must also be considered.

The output from the lock-in amplifier is sampled a number of times by the computer's ADC. The average of these samples is used as the measure of the thermoelastic signal. When the mirrors of the scan system are moved, the detector 'sees' photons from the whole path along which the mirrors are moved. If the motion of the mirrors is large this is an important factor, but generally the mirror motions from one point to the next are quite small and the thermal change is small as well. During the mirror movement time, the lock-in amplifier output is not sampled by the computer.

To give the reader an idea of the time scale involved in mirror motions I shall mention that for small motions it takes the mirrors on the SPATE 9000 at least 8 ms to move to within 95% of the new location. For larger motions it can take up to 15 ms.

Once the mirrors do have the detector pointed at the right spot for a measurement it will take some time for the lock-in to produce a measurement. For example it takes a two-pole low-pass filter 6.6 time constants to settle to within 99% of a new final value. If the time constant, $T$, is set too long in relation to the mirror scanning rate, a portion of the previous reading will be dragged into the new reading. Because of this it is a good idea to set the scan rate so that the time constant of the lock-in is 10 to 20 times shorter than the time the mirrors spend at any single position. A factor of 10 here keeps the amount of corruption of a reading from the previous reading to below 10% of the difference between the two.

The factor of 10 between the lock-in time constant and the time per scan position can leave the operator with a noise bandwidth that is large enough that the twice reference frequency component of the mixer output will not be sufficiently attenuated by the lock-in's low-pass filter. To handle this, the computer of the SPATE 9000 system is programmed to sample the output of the lock-in for an integral multiple of one-half the period of the reference signal. The SPATE 9000 samples the output of each channel of the lock-in 45 times during the sampling time.

The digital sampling of the computer can further eliminate noise from the detector/lock-in signal. The noise bandwidth for sampling is

$$f_n = 0.5/S \qquad (3.2)$$

where $S$ is the sampling time. Notice by comparing equations (3.1) and (3.2) that the lock-in amplifier is a much more effective means of filtering than using the sampling time. But because the time constant must be set to less than 10% of the sampling time, the smoothing produced by averaging many samples taken during the sampling time is important to noise bandwidth reduction. Notice that, unlike the low-pass filter integration, the computer averaging (integration) does not carry a residual from the previous reading.

A typical example of acquisition parameters and the resulting noise bandwidth is shown below.

Specimen loading:   10 Hz
Sampling time:   0.1 s

Lock-in time constant:   0.01 s

Noise bandwidth of lock-in output:   12.5 Hz
Noise bandwidth of sampled data:   5.0 Hz

If the operator chooses to scan the specimen twice and average the two scans, then the total sample time would be 0.2 s per pixel. The resulting noise bandwidth of the sampled data would be 2.5 Hz.

It is important to point out that the noise bandwidth of the lock-in and the noise bandwidth of sampling do not add. The final noise bandwidth of the data is the smaller of the noise bandwidth of the sampling and of the lock-in time constant.

During the above discussions, the equivalent noise bandwidth has been used to help explain the noise that will appear on the output of the lock-in and consequently in the TSA data. To help understand how much noise can be expected in the TSA data it is important to know that the noise reduction is proportional to the square root of the noise bandwidth. This means that if the noise bandwidth is improved by a factor of 4 (by say increasing the sampling time by a factor of 4), then the noise reduction will improve by a factor of 2.

### 3.2.4 Special signal processing considerations

There are many difficulties in acquiring thermoelastic data that have not been discussed up to this point. These problems generally arise as the result of trying to meet the needs of a particular stress analysis. Many of these difficulties have been conquered, and in this section a few of the successes are pointed out and discussed briefly.

#### 3.2.4.1 Problems caused by mirror movements

When mirror movements are large, the view of the IR detector is swung over a long path. This is sometimes the case when complicated specimen geometries are scanned and the computer has been programmed to ignore unimportant areas. These large mirror motions can cause two problems, erroneous signals leading to lock-in overload, and detector current bias being out of range. The erroneous signals are the result of variations in photon emission being picked up along the path that the detector sees as the mirrors swing to a new position. The signals produced as a result of these variations may produce lock-in inputs or outputs large enough to cause overloads. This problem can be handled in a number of ways. The user can:

(i)  set-up a geometry that has large swings but starts sampling again in an uninteresting area, so that by the time scanning moves onto an interesting stress, the electronics have settled back down;

(ii) provide a background of even temperature for scanning paths beyond the specimen surface; and/or

(iii) set the computer to monitor and handle lock-in overloads, by waiting for signals to settle.

Another more serious problem caused by large mirror movements is a problem with the detector bias current. The detector is a semiconductor junction that has a constant current passing through it. The amount of current is dependent upon the heat flux reaching the detector. When the detector is quickly swung onto a surface of different temperature, the current must be adjusted. This adjustment has a time constant in the neighbourhood of seconds. Therefore, a settling time is required when the detector's view moves onto a surface of different temperature. This problem is generally handled the same way as the lock-in overload problem mentioned above.

### 3.2.4.2 Noisy load signals

The lock-in amplifier can be used to measure stress amplitudes even if the specimen loading is not single-frequency cyclic loading. In this case the detector signal will carry frequency components that are the result of loads that are not at the frequency of the reference signal. The lock-in treats these signals just as it would any other noise.

In Chapter 6 a much different approach to detector signal processing is discussed. In that chapter the use of a fast Fourier transform (FFT) as a replacement for the lock-in amplifier signal processing is explained. Frequency-domain signal analysis provides a number of important capabilities to TSA.

### 3.2.4.3 Tests at elevated temperatures

When the temperature of the target is higher than room temperature, the thermoelastic emissions are larger (see equation (1.3) in Chapter 1). The detector's response is proportional to the amount of heat flux that is incident upon it, and this is proportional to the third power of the temperature. A good reference on this whole topic is Enke (1989). The large amount of heat flux must be attenuated or it will do damage to the detector of the SPATE 9000. This is done by using filters. Filters are also important for limiting chromatic aberration. Analysis at elevated temperatures may produce other problems, such as thermal gradients and changes in emissivity over the surface of the specimen. A partially transparent optical chopping technique is under development by Lesniak at the University of Wisconsin that should solve many problems of high-temperature applications of TSA. A standard optical chopper has also been used (Saul and Webber 1987) to attack these problems by measuring the absolute or DC thermal emissions from the specimen surface. By using filtering and chopping methods, TSA scans at temperatures as high as 1100°C have been made.

### 3.2.4.4 Specimen motion

Specimen motion caused by the cyclic loading needed for TSA can be significant enough to cause problems. A specimen edge moving across the measurement spot, or specimen motions large enough to blur the TSA image are the two major problems. Commonly the most effective method for handling these problems is to use a viewing angle that minimises the amount of the specimen motion perpendicular to the line of sight of the IR detector. Sometimes a mirror connected to, and moving with, the specimen can be a great help in getting a good line of sight. Another solution is to use a device that Ometron offers which moves the camera scan mirrors through a phased and scaled sinusoidal pattern that is synchronous to the reference signal.

## 3.3 SOFTWARE

The TSA technology places heavy demands on computer-based handling of the acquired data. Already we have discussed most of the major problems involved in acquiring thermoelastic data and how they have been met through the design of the TSA instrument and the acquisition software. In this section we will concentrate on the stress analyst's needs for thermoelastic data storage, manipulation, and presentation.

### 3.3.1 Data storage

Thermoelastic data are generally collected from a rectangular matrix of measurement positions. Consequently, the data are stored in the computer's memory as a matrix whose dimensions represent the physical size of the scanned area, and whose values represent the amplitude of heat flux oscillations. Commonly there are two planes or matrices of data, the $X$- and the $Y$-planes. The $X$- and $Y$-values of a pixel can be combined to give the amplitude and phase of the thermoelastic signal.

As was demonstrated in section 3.2.1, a TSA system generates a large amount of data. A most important responsibility of any TSA system is to provide competent storage and recall of data. However, merely being able to save and recall a data file is insufficient. The data structure itself must be competent or else the stored data will soon crumble to uselessness. If a TSA data set is saved to disk and it does not include all the important information about the way the data was collected (cyclic load amplitude, specimen number, test date, operator, load frequency,...) then soon human memory will fade and the data will be useless. Another way to make data useless is to limit accessibility to it. SPATE 9000 data sets are easily movable (portable) to other computers. This helps take advantage of advances in computer and data analysis technology.

The SPATE 9000 data can be easily documented, saved, catalogued, searched, archived, and retrieved. This allows the stress analyst to turn his attention to analysis of the thermoelastic data. For this task he will need data presentation and data manipulation tools.

### 3.3.2 Data presentation

The SPATE 9000 provides a number of presentation tools and operations to aid in the collection of thermoelastic data. But after the data is collected the stress analyst needs tools that will help to quickly identify specific areas of interest. These needs are centred around the requirements for speed and data comparison. Later, when the analyst has located the interesting features of the data, he wants to have presentations that will help him communicate ideas and results to his associates.

To provide a quick yet accurate display of the thermoelastic data, a display of the data as many small colour-filled rectangles with colours keyed to stress level is used. Constructing this graphic, or any other, can take anywhere from seconds to minutes, depending on the available computing power. To make sure time is not wasted each time an image is completed, the computer pushes a copy of it into a screen stack. The stack holds the last five screen displays presented by the computer. The operator can recall any of the displays from the stack through the *Screen Memory* feature of the programme. Another tool that is provided to aid comparison between data sets is a split-screen feature. The operator can use a split-screen mode to put two graphic presentations on the screen side by side.

The SPATE 9000 software uses the screen stack to implement a type of windowing display environment. Each time a page of textual information is presented, such as scan parameters or graphics set-up parameters, the current graphics are pushed onto the stack so that they can be restored when the textual operations are finished. This ensures that a graphical presentation of thermoelastic data is kept on the display whenever the display is not needed for text.

The spatial nature of the TSA data produces a number of special presentation requirements. Since the TSA image is collected with a 'slow scan' mirror system targeted by a visible light channel, there is a need to be able to interrogate the data with the mirrors moving in co-ordination with a display cursor. In the SPATE 9000 the cursor can be used to mark out a path on the scan area along which the data values can be plotted. Figure 3.3 shows a TSA data set that has been interrogated in this way and presented through the split-screen feature.

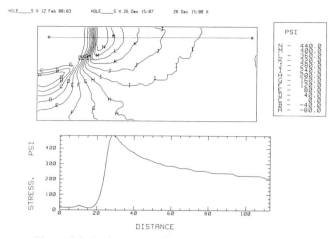

**Figure 3.3** An interrogated contour plot of TSA data.

Figure 3.3 hints at some of the versatility of the presentations that are available in the SPATE 9000 system. A diversity of data presentation types is important to the stress analyst or researcher for clear communication of test results. The requirements of data visualisation and reproducibility often dictate a certain type of graphic. The SPATE 9000 is equipped with the ability to produce the contour plots, line plots, and colour-filled grid displays that have already been mentioned. In addition, the 9000 provides colour contour plots, colour-filled contour plots, and isometric or mountain plots of the thermoelastic data. Additionally, facilities are provided to the user to add his own data visualisation code.

### 3.3.3 Data manipulation

To characterise, quantify, and enhance the collected data is an area of much interest to most TSA users. There is an almost unending list of useful processes that can be imposed on TSA images. Most of them are closely tied to the way that the thermoelastic data are collected or presented. A good example of this is the smoothing functions provided by the SPATE 9000.

The two types of smoothing commonly applied to thermoelastic data are binomial convolution and median filtering. Both smoothing methods modify a data pixel based on the magnitude of its neighbours in the data matrix. The binomial convolution replaces each pixel with the weighted average of itself and its eight nearest-neighbours. The weighting scheme is indicated in the following 3 by 3 matrix:

$$\begin{matrix} 0.0625 & 0.125 & 0.0625 \\ 0.125 & 0.25 & 0.125 \\ 0.0625 & 0.125 & 0.0625 \end{matrix}$$

The median filter replaces each pixel with the median of the data from each of the nine locations in the 3 by 3 neighbourhood of the target pixel. The result of both of these techniques is that spatial information is used to eliminate noise from the thermoelastic image.

The median filter has an advantage over the binomial convolution in that it tends to cause less distortion of the data. A binomial convolution may make an image look blurred. The algorithms currently used for these smoothing operations take significantly different amounts of time. The median filter with its need to sort the neighbourhood data takes the longest.

When thermoelastic data are particularly noisy, it is common to smooth the data more than once. The smoothing is most often done to aid in presentation of the results of TSA work. The goal of the smoothed presentation is often to provide clear communication of results to designers or the confirmation of FE results.

Another useful type of data manipulation is the statistical analysis of scan data. The statistical analysis algorithm of the SPATE 9000 generates the minimum, maximum, mean, standard deviation, and amplitude distribution of the measured stresses (e.g. figure 3.4). The 9000 presents a plot of the percentage of pixels at each stress level. This tool has found wide usage: from determining nominal stress level, to helping quantify damage in composites (Zhang and Sandor 1989).

The collection of the $X$- and $Y$-components of the detector signal allows for the calculation of amplitude and phase information. A presentation of the $X$- and $Y$-data, or of the phase data, can help determine the quality of a TSA scan. As was pointed out in earlier chapters, if adiabatic conditions are not maintained, the thermoelastic formulae do not hold. When stress gradients are high, so are thermal gradients. Large thermal gradients combined with low rates of loading can result in enough conduction to effectively invalidate the adiabatic assumption. A sign that conduction is a

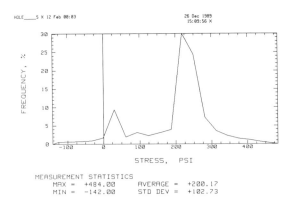

**Figure 3.4** Statistical analysis of the thermoelastic data.

factor in a scan is the presence of a phase shift near high-stress gradients (Lesniak 1988). Because of this, a plot of the $Y$-component or phase of a scan can be used to indicate if the stress analyst should be concerned about heat conduction.

The SPATE 9000 offers its user the ability to calculate and present the amplitude and phase images as well as the ability to 'correct' the phase of the data and recalculate the $X$- and $Y$-components of the TSA image.

Algebraic operations applied to whole TSA scan images are also available on the SPATE 9000. Absolute value, addition and subtraction are built into the 9000 and can be accessed from function keys. These operations are currently being used to aid in comparison between scans, or for high-temperature correction schemes in work being undertaken by Lesniak at the University of Wisconsin.

Research users of TSA systems often conceive of new ways of using TSA to advantage. For these people it is important that the SPATE 9000 has enough flexibility to be easily used as their research instrument. With this in mind the 9000 has been designed with facilities and a software architecture that provides a user with the ability to modify the way the 9000 operates. The main body of the SPATE 9000 code is open to the user for modification. Along with this, documentation is provided about important variables and sub-routine operation.

To help provide further access to TSA technology, the SPATE 9000 software is written in Hewlett Packard Basic. Because Hewlett Packard sells a broad line of computers that all run the SPATE 9000 software well, the right level of price and performance is available for a broad range of budgets at locations all over the world. The hope is that advances to TSA technology can be quickly implemented throughout the TSA community merely through the dissemination of software.

Recent advances have improved the productivity of TSA for commercial environments. Productivity for these applications is critical if the technology is to survive economically.

# 4

# Signal Attenuation Due to Internal and External Factors

## J McKelvie and A K MacKenzie

## 4.1 INTRODUCTION: NON-ADIABATIC EFFECTS

The use of the simple connection between the amplitude of the temperature fluctuation, $\delta T$, and that of the stress fluctuation, $\delta\sigma$, is dependent upon the presumption of adiabaticity, that is, that all of the heat generated in an element goes towards producing a temperature rise in that element, and does not leak away into, for example, surrounding material or the environment. If this condition of adiabaticity is not maintained, then meaningful analysis of the signal generated by the sensor becomes more problematical, and it is therefore important to have some means of assessing the degree to which the corresponding error is significant in any particular case (such an error will always exist; what concerns us is its magnitude).

Belgen (1968) was the first to address this matter, identifying three distinct mechanisms by which these errors would arise: firstly, heat loss from the element to the environment by conduction, convection, and radiation, from the surface being observed; secondly, conductive heat loss from the element to surrounding material whose temperature is less than that of the element under consideration (and vice versa, there may be heat gain from surrounding material); thirdly, in those cases where, for the purposes of improving the level or the uniformity of radiation, a coating is applied that has a significantly different thermoelastic constant (on a strain rather than stress basis) from the testpiece, then the temperature amplitude on the external surface will in general be different from that at the surface of the test material.

## 4.2 THEORETICAL ANALYSIS

Subsequent to Belgen's work, more rigorous contributions have been made by McKelvie (1987) and MacKenzie (1989), and it is these later papers that provide the basis for much of this chapter. The analyses are not presented in detail, but are based upon solutions of the heat conduction equation

$$k\nabla^2 T = \rho C_p \frac{\partial T}{\partial t} - \dot{Q}(x, y, z, t) \tag{4.1}$$

where $\dot{Q}$ is the rate of heat production per unit volume within the material. The various cases considered all have $\dot{Q}$ as a sinusoidal function of time, and different geometries and stress distributions are represented through Fourier series expressions and appropriate boundary conditions.

## 4.3 HEAT LOSS TO THE ENVIRONMENT

### 4.3.1 Analysis

Belgen (1968) presented an analysis of a plate with internal sinusoidal heat production and cooled by convection on its two surfaces. McKelvie (1987) presents a more exact analysis based upon a solution of equation (4.1) for an infinite slab lying between $-l < x < l$. For a rate of heat production $A_0 \sin(\omega t)$ that is constant throughout the thickness (i.e. modelling a state of uniform stress), it is shown that the ratio of the non-adiabatic temperature amplitude to the adiabatic is given, to a good approximation, by

$$\frac{T}{T_{ad}} = \sum_{n=1}^{\infty} \frac{2}{[(x^2 \alpha^4{}_n/\omega^2) + 1]^{1/2}\{l[(\alpha^2{}_n/H') + H'] + 1\}} \tag{4.2}$$

where $H' = H/k$, $H$ is the surface heat transfer coefficient ($J\,s^{-1}\,m^{-2}\,K^{-1}$) and $\alpha_n$ are successive positive solutions of the equation $\alpha \tan(\alpha l) = H'$. Figure 4.1 shows this attenuation factor plotted for two materials for several values of $H$. A value of $H = 10$ corresponds to natural convection at room temperature; $H = 100$ represents natural convection at 1000 K or moderate forced convection at ambient (e.g. normal atmospheric) winds; and $H = 1000$ is typical of severe forced convection at ambient conditions such as in high-speed wind-tunnel conditions.

### 4.3.2 Practical significance

It is clear that for titanium the only conditions that would give grounds for concern at frequencies above 1 Hz would involve very thin material; more conductive materials would be even less affected.

The behaviour illustrated in figure 4.1 corresponds to a simple qualitative consideration, which would indicate a greater attenuation with increasing $H$, and with decreasing frequency and conductivity. With the PVC, it can be seen that significant attenuation may occur, even in substantial thickness, at frequencies of several hertz in the moderate convection case. Since the local surface temperature will depend upon the local $H$-value, and since this latter will vary across the surface (e.g. in rebates as compared with high spots), this effect cannot be 'calibrated out', and we must be aware of possible errors arising from it under appropriate conditions.

## 4.4 CONDUCTIVE HEAT LOSS TO SURROUNDING MATERIAL

### 4.4.1 General analysis

Belgen's analysis of internal condition was rather simplistic, considering only the case of a stressed element close to a relatively massive unstressed region. The analysis of McKelvie (1987) is more general and gives the solution for a one-dimensional sinusoidal spatial distribution across an infinite plate of width $4l$ in the $x$-direction, and again sinusoidal in time, as

$$\frac{T}{T_{ad}} = \frac{\omega}{\sqrt{\omega^2 + K^2}} \qquad (4.3)$$

where $T$ is the amplitude of the temperature change, and

$$K = \frac{\chi \pi^2 a^2}{4l^2} \qquad (4.4)$$

and $a$ indicates the spatial frequency, the rate of heat production being given by

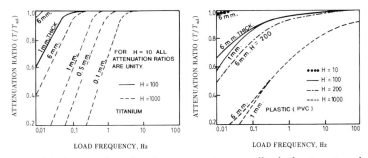

**Figure 4.1** Attenuation of surface temperature amplitude due to external heat losses (versus loading frequency) for two materials and their relevant thicknesses. $H$ = heat transfer coefficient ($J\,s^{-1}\,m^{-2}\,K^{-1}$).

$$A(x, t) = A_0 \cos\left(\frac{a\pi x}{2l}\right) \cos \omega t. \tag{4.5}$$

Moreover, the actual temperature lags that of the adiabatic condition by a phase angle

$$\theta = \tan^{-1}(K/\omega). \tag{4.6}$$

Figure 4.2 plots the factor $(T/T_{ad})$ of equation (4.3) for three materials. As expected the more conductive materials are more affected, it being easier for the heat to 'escape' sideways in such materials. Interestingly, the higher the spatial frequency, the greater is the attenuation; this concurs with intuition, since the temperature gradients—and therefore the rate of heat transfer—are larger with higher frequencies (amplitude remaining constant). High load-cycle frequency operation minimises the effect, as even the simplest considera-tion would indicate.

### 4.4.2  Relevance of the analysis

Possibly the most significant application of this analysis is the consideration of the consequences in trying to determine the details of small-scale features—such as around the tip of a crack—for wherever such detail is of importance, it implies that there are high spatial-frequency terms and therefore errors will be automatic, unless we can operate at appropriately

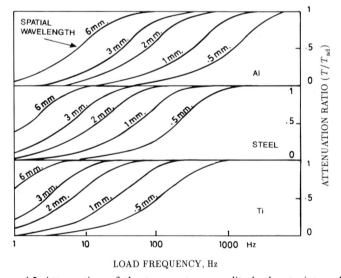

**Figure 4.2** Attenuation of the temperature amplitude due to internal conduction (versus loading frequency) for three materials and various one-dimensional spatial wavelengths.

high load frequency or introduce a system of compensating the relevant spatial frequencies with appropriate amplification so as to negate the attenuation.

The matter of the phase change is also relevant; the signal processing techniques typically used for SPATE are based on correlation with a reference signal derived from the load applied to, or strain induced in, the structure. Such techniques give a basis for a measure of the phase of the observed IR response relative to the stress or strain in the structure, but care should be taken to correct for phase shifts induced either by virtue of rapid variations in the stress distribution (as indicated above), or by surface coatings (discussed later in this chapter), or by instrumentation. In addition, the software supplied with the SPATE 8000 system assumes a constant phase relationship between the reference and the response. Automatic logging of variations in phase across the structure is not provided for. In the case of the current SPATE 9000 system there is commercially available software which caters for logging and processing of phase information (see Chapter 3).

### 4.4.3 2D and 3D effects

#### 4.4.3.1 2D

McKelvie (1987) showed that if we have a two-dimensional sinusoidal distribution of stress, giving a heat generation rate

$$A(x, y, t) = A_0 \cos\left(\frac{a\pi x}{2l}\right) \cos\left(\frac{b\pi x}{2l}\right) \cos \omega t \tag{4.7}$$

then the solution is given by equation (4.3) except that $K$ is modified to

$$K_{xy} = \frac{\chi^2 \pi^2 (a^2 + b^2)}{4l^2}. \tag{4.8}$$

This means a more severe attenuation, and a bigger phase lag. This again confirms expectations, since heat is now free to flow from hot spots in any direction in the plane, rather than just one.

#### 4.4.3.2 3D

The effect of a 3D distribution

$$A(x, y, z, t) = A_0 \cos\left(\frac{a\pi x}{2l}\right) \cos\left(\frac{b\pi y}{2l}\right) \cos\left(\frac{c\pi y}{2l}\right) \cos \omega t \tag{4.9}$$

is again analogous to the 1D case, except that $K$ becomes

$$K_{xyz} = \frac{\chi^2 \pi^2 (a^2 + b^2 + c^2)}{4l^2} \tag{4.10}$$

giving yet more attenuation and phase lag.

### 4.4.4 A specific case: the plate in bending

Having a description of the temperature response for a sinusoidally distributed stress pattern, it becomes possible to describe the response due to any stress pattern, using its Fourier composition. An example is the plate in bending, for which the stress distribution is linear. McKelvie (1987) uses the Fourier representation of a periodic triangular wave-form to describe this case, and gives the result for the attenuation at the most highly stressed points in a plate of thickness $2l$, as follows

$$\frac{T}{T_{ad}}=\frac{8}{\pi^2}\left[\left(\sum_{n=0}^{\infty}\frac{\omega^2}{(K_n^2+\omega^2)(2n+1)^2}\right)^2+\frac{1}{\omega^2}\left(\sum_{n=0}^{\infty}\frac{K_n\omega^2}{(K_n^2+\omega^2)(2n+1)^2}\right)^2\right]^{1/2}$$

(4.11)

where

$$K_n=\chi^2\pi^2n^2/4l^2.$$

(4.12)

Again, there is a phase lag in the response. It arises out of the additions of all the phase-shifted spatial frequencies and has no simple expression, the respective phase angles being

$$\theta_n=\tan^{-1}(K_n/\omega).$$

(4.13)

In general, the resultant phase shift will be greater than that of the fundamental ($n=1$). (Equation (4.11) for the attenuation includes the effect of the phase angles in the summation.) Figure 4.3 shows equation (4.11)

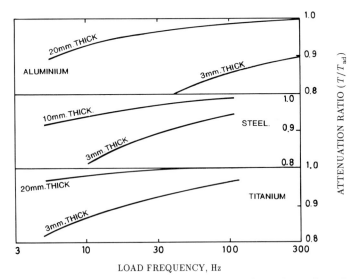

**Figure 4.3** Attenuation of the temperature amplitude on the surface of a plate in bending (versus loading frequency) for three materials and various thicknesses.

plotted for specimens of three metals of various thicknesses. The expected behaviour, whereby increasing thermal diffusivity and decreasing thickness (i.e. steeper thermal gradient) cause more severe attenuation, is obvious.

### 4.4.5 Some experimental corroboration

Belgen (1968) and Dillon and Taucher (1966) obtained experimental results for bars in bending, using radiometric methods and thermocouples, respectively. Figure 4.4 compares their findings with equation (4.11), and the correspondence is seen to be rather good. This indicates that the analysis thus far gives a reasonable description of the physics. The work of Beghi *et al* (1987) confirms the phase-shift effect in a non-uniform stress field.

### 4.4.6 Practical significance

Several points of importance arise: firstly, wherever we have a varying stress sum condition, the amplitude of the temperature variation at any point will in general be less than it would be for the same stress sum amplitude in a uniform field (all other things being equal), the attenuation being worse in cases of sharp maxima of either sign (it is in fact possible under certain combinations of spatial distribution, material properties, and loading frequency, to get an amplification in some regions rather than an attenuation; this occurs as a result of an interaction between the rate of heat generation

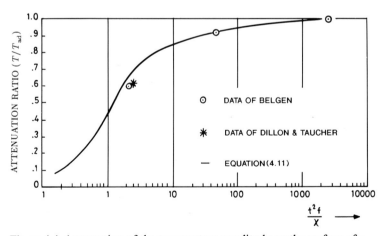

**Figure 4.4** Attenuation of the temperature amplitude on the surface of a plate in bending (versus a non-dimensional parameter). Comparison of equation (4.11) with experimental data.

and the rate at which thermal waves travel across the field—similar to a resonant condition; one example is indicated by Belgen (1968) in the treatment of lateral conduction); secondly, the estimate of the response amplitude may be further reduced due to analytical error if the phase lag which accompanies attenuation is neglected (e.g. if the phase of the response with respect to the load or strain reference is assumed to be constant across the structure); thirdly, as we seek more and more detail of a stress sum distribution we run into increasing difficulty, because what we are in effect trying to do is to elicit the higher spatial-frequency content, and those higher frequencies are more severely attenuated; fourthly, the detection of a phase lag in a uniform surface stress field indicates that there is a stress sum variation in the thickness of the field specimen—indeed it may be possible to use the phase lag to determine quantitatively the stress sum gradient at the surface into the interior; fifthly, while it is possible in principle, knowing the thermal properties of the material and the load-cycle frequency, to correct any given apparent distribution by Fourier transforming and correcting the various components in amplitude and phase, this requires some knowledge of what is happening in the thickness; lastly, all of this pre-supposes that there are no attenuations of phase shifts due to other effects. An important instance where such other effects occur arises with the use of surface coatings, which will now be considered.

## 4.5 THE EFFECTS OF SURFACE COATINGS

### 4.5.1 The nature of the problem

In the measurement of the thermoelastic response by radiometric means—as in SPATE—it transpires that for many specimen materials it is beneficial to apply a high-emissivity coating to the surface. This has two effects: firstly, the radiation is enhanced, so that the signal level is maximised; secondly (and in principle more importantly), the emissivity is then known to be fairly uniform. Without reasonable uniformity of emissivity, radiometric stress measurement becomes impossible unless of course the emissivity pattern can be established separately. The use of these coatings is especially recommended for metallic specimens, since the emissivity is typically both low and somewhat variable, being dependent upon the local surface state (reflectivity, roughness, degree of oxidation, etc.).

The usual high-emissivity coatings recommended are black plastic-based paints (usually carbon-loaded) (MacKenzie 1989). It is unfortunate that precisely in those cases where it is most required—the metallic specimens—there is a considerable mismatch in the strain-based thermoelastic constants of coating and substrate (it will be clear that the coating and substrate experience the same strain). The coating is substantially thermoelastically inert in comparison with the substrate, and this may give rise to undesirable effects.

It will be appreciated readily that such a coating will tend to act as an insulating layer, and that the surface temperature of the coating will alter due to variations in its thickness. Less obvious at first is the effect whereby the coating will act as a thermal inertia that requires heat to be supplied from the substrate, which, in turn, must therefore experience a reduction in its temperature fluctuation. If the substrate is of very high thermal diffusivity, then there will be little effect due to this mechanism, because heat may then be drawn from the bulk of the material thus inducing little suppression of the surface temperature; in very low diffusivity substrates, on the other hand, the heat to drive the coating's temperature rise must come in the main from the substrate's surface region, which may therefore experience greater attenuation. McKelvie has called these two mechanisms 'thermal lag' and 'thermal drag-down'. The question is whether they are significant effects when using the kinds of paints (and the typical thicknesses of paint) that are likely to be applied.

### 4.5.2. Analysis: thermal lag

Three analyses have been presented. The first, by Belgen (1968), considered the coating as having a simple exponential response to changes in substrate temperature. The second, more exact, analysis of McKelvie (1987) modelled the coating as half of an infinite slab having on both faces temperature profiles that vary sinusoidally and synchronously in time (by symmetry, there can be no heat flux across the central plane of such a slab, which plane may therefore represent the outer surface of the paint when the heat loss to the environment is negligible). The third analysis, in MacKenzie (1989), considers the heat flux in terms of the thermal waves that sweep across the slab under the conditions of oscillating temperature profiles at one side. This, the most comprehensive model, reduces to the second model when the waves are considered to reflect 100% with 180° phase inversion at the substrate interface and 100% with 0° phase inversion at the air interface. The wave model is more flexible in that different boundary conditions may be more readily applied, and, at the time of writing, work continues to quantify such modified boundary conditions and consequent modified interference effects which will arise with implications for possible optimising of the substrate interface condition to encourage heat flow into the coating (analogous to graded refractive index systems, or anti-reflection coatings, in optical technology).

The model of McKelvie (1987) gives the result for the temperature on the paint surface as

$$\frac{T_o}{T_i} = \left| \frac{1}{\cosh \mu t_c (1 + i)} \right| \tag{4.14}$$

where $\mu = (\omega/2\chi)^{1/2}$, $t_c$ is the thickness of the coating, $T_o$ is the outer surface temperature amplitude, and $T_i$ is the inner surface temperature amplitude.

There is also a phase lag at the outer surface given by

$$\theta = \arg \left( \frac{1}{\cosh \mu t_c (1 + i)} \right). \tag{4.15}$$

The questions that now arise are what thermal diffusivity and what thickness of paint are appropriate. Belgen (1968) indicates that paint coatings should be at least 10 microns thick in order to avoid transparency (and therefore variation in emissivity), and the work of MacKenzie (1989) suggests a minimum of 20 microns, as will be discussed. Unpublished trials by the first author indicate that a spraying process may, in unfavourable situations such as curves and rebates, result in thicknesses exceeding 40 microns when trying to produce a 15 micron coating. In the matter of thermal diffusivity, Belgen measured one carbon-filled acrylic at $0.5 \, \text{m}^2 \, \text{s}^{-1}$. Subsequent experience reported by MacKenzie (1989) indicates that his value may be high by a factor of three or four for the paints commonly recommended. Figure 4.5 shows equation (4.14) plotted to a frequency abcissa for various relevant thicknesses and $\chi = 0.5$.

### 4.5.3 Analysis: thermal drag-down

An analysis of this effect has been reported by McKelvie (1987). The model is rather a simple one, in that the paint coating is represented by a layer of

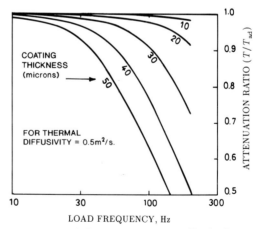

**Figure 4.5** Attenuation of the temperature amplitude due to coating thermal lag (versus loading frequency) for various coating thicknesses and one thermal diffusivity value $\chi = 0.5$.

substrate in which the rate of heat production is zero, with the paint thickness adjusted in the ratio of the volumetric heat capacities—so that the inert substrate layer has the same heat capacity per unit surface area as the actual paint layer would have. The matter of thermal lag through the inert layer is omitted in this model. Instead, a choice is made as to where the attenuation is to be calculated, for example between the paint surface and the paint—substrate interface, such that the calculation will underestimate the attenuation. (This choice is based on whether the substrate is more or less conductive than the paint.)

The solution given for the response to an oscillating heat production of constant amplitude through the substrate thickness is

$$T(z,t) = \frac{4A_0}{\pi \rho C_p} \sum_{n=1}^{\infty} \frac{\cos(n\pi z/2l)}{n(K_n^2 + \omega^2)^{1/2}} \sin(\omega t + \theta_n) \left( \sin\frac{n\pi}{2} - \sin\frac{n\pi d}{2l} \right) \quad (4.16)$$

where

$$K_n = \chi\pi^2(2n-1)^2/4l^2 \quad (4.17)$$

$$\theta_n = \tan^{-1}(K_n/\omega) \quad (4.18)$$

$z$ is the coordinate in the thickness direction, $l$ is the plate dimension in the $z$-direction, to be taken as very large in comparison with the paint thickness (this analysis is invalid for a thin plate), and $d$ is the thickness of the inert layer. Each component of the summation has a different phase angle, which makes the summation more complicated. Ignoring this effect so that, again, the attenuation is underestimated, the ratio of the amplitudes is

$$\frac{T(z)}{T_{\mathrm{ad}}} = \frac{4}{\pi} \sum_{n=1}^{\infty} \frac{\omega \cos(n\pi z/2l)}{n(K_n^2 + \omega^2)^{1/2}} \left( \sin\frac{n\pi}{2} - \sin\frac{n\pi d}{2l} \right) \quad (4.19)$$

where $T_{\mathrm{ad}}$ is the temperature that the surface would reach with no inert layer. The evaluation is done at $z=0$ (outer surface) for conductive, and $z=d$ (the interface) for a non-conductive substrate.

The particular case illustrated in McKelvie (1987) is that of a plate in bending. The corresponding result is

$$\frac{T(z)}{T_{\mathrm{ad}}} = \frac{4}{\pi} \sum_{n=1}^{\infty} \frac{\omega \cos(n\pi z/2l)}{n(K_n^2 + \omega^2)^{1/2}}$$

$$\times \left( \frac{2l}{n\pi} \sin\frac{n\pi(l-d)}{2l} - (l-d)\cos\frac{n\pi(l-d)}{2l} \right) \sin\left( \frac{n\pi}{2} \frac{1}{l-d} \right). \quad (4.20)$$

This relation is shown in figure 4.6 for the same substrates as in figure 4.3. (Non-metallic substrates are not included as these would, in general, have thermoelastic constants such that the paint could not be considered inert;

furthermore, in non-metals—as previously indicated—emissivity enhancement may well be unnecessary.) A worsening effect with decreasing conductivity is obvious from the plot.

### 4.5.4  Combined thermal lag and drag-down

McKelvie (1987) suggests that, as a first approximation, the attenuation factors as calculated for lag and drag-down be simply multiplied to give an estimate of the combination of the two separate effects.

### 4.5.5  Experimental corroboration

*4.5.5.1  Low-frequency studies*

The authors have carried out some practical investigations of these effects, and this work is reported here for the first time.

Plain 'dog-bone' specimens 12.7 mm thick and 50 mm wide were prepared with four strips of paint coatings of different thicknesses, from approximately 15 microns to approximately 50 microns. (The thickness was measured using an inductive probe.) The specimens were of different

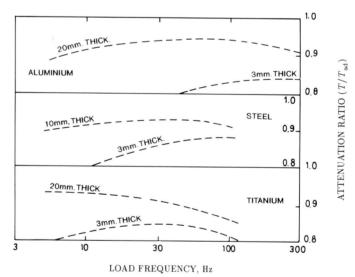

**Figure 4.6** Attenuation of the temperature amplitude due to the combined effects of bending and coating thermal drag-down (versus loading frequency) for three materials and various thicknesses; to be compared with figure 4.3.

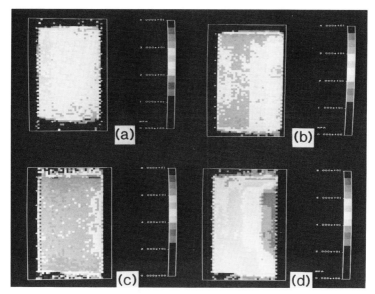

**Figure 4.7** SPATE patterns on uniformly stressed specimens having vertical strips of paint of different thicknesses: (*a*) copper specimen at 5 Hz, (*b*) copper specimen at 25 Hz, (*c*) stainless steel specimen at 5 Hz, and (*d*) stainless steel specimen at 25 Hz.

conductivity—copper, mild steel, and stainless steel—and were cycled at a variety of frequencies up to 30 Hz. The tests were carried out under load control with the output from the load cell providing the reference signal for the correlator. Some of the results are shown in figures 4.7 and 4.8 and table 4.1. They confirm a number of the predictions of the above analysis:

(i) On each substrate, signal attenuation is noticeable as the thickness increases except that in the case of copper an increase occurs as we go from the thinnest coating strip to the next—this is ascribed to an inadequate thickness of the paint (only 10 microns average in the thinnest case), leaving it somewhat transparent to the 8 to 12 micron radiation;

(ii) the attenuations become worse with increasing frequency;

(iii) the less conductive substrate exhibits the effect to a greater degree; and

(iv) there are significant phase shifts in the signals due to the coatings— these also increase with increasing paint thickness and with frequency.

In interpreting figures 4.7 and 4.8 it should be noted that, due to the way the measured data were logged and displayed, the data presented do not strictly represent the amplitude of the measured response. In this instance for each substrate material at each load-cycle frequency the zero-phase reference

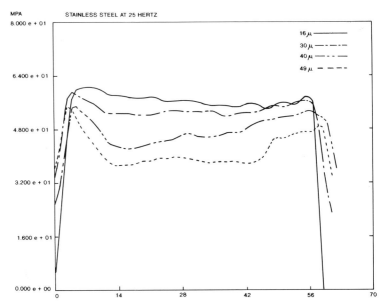

**Figure 4.8** SPATE $\delta\sigma$ values on sections through the four strips of figure 4.7(d).

was based on the response signal from the coating thickness giving the greatest measured response. The data logged and presented here is the component of the measured response in phase with this zero-phase reference. For the coating bands of other than optimum thickness this gives an underestimate if used as a direct measure of the response amplitude. The corresponding relative phase information may be used to apply a simple numerical correction.

As in the case of conduction effects in the substrate, information relating to the variation in phase across the structure of the measured response gives a

**Table 4.1** Relative phase lag for RS matt black paint on a copper substrate.

| Frequency (Hz) | Paint thickness (microns) | | | |
|---|---|---|---|---|
| | 10 | 22 | 32 | 45 |
| 5 | 0° | 2° | 3° | 6° |
| 10 | 7° | 10° | 13° | 20° |
| 20 | 12° | 17° | 23° | 34° |
| 30 | 24° | 30° | 39° | 51° |

basis for detecting the presence of variations in attenuation due to thermal 'lag' (e.g. due to variations in coating thickness) and also contributes to a basis for estimating the magnitude of such effects with a view to correcting for them.

In these tests no attempt was made to determine quantitatively what proportions of the attenuation were due to the separate 'lag' and 'drag' effects, but it is clear that both are contributing in the case of stainless steel.

It will be noticed that in all cases the attenuation appears reduced at each end of the paint strip but, at present, no definitive explanation can be offered.

### 4.5.5.2 High-frequency studies

MacKenzie (1989) reports selected results of a more extensive study of coating effects up to 300 Hz. Figures 4.9 and 4.10 illustrate the nature of the findings, which, again, confirm the general behaviour predicted. Several proprietary paint formulations were investigated. All showed the same qualitative features in their response data.

An interesting feature that appeared consistently was a recovery of the signal seen in the region beyond about 50 Hz at thicknesses above 70 microns. At present this is being attributed to the thermoelastic effect of the coating itself and constructive or destructive interference with the 'main' effect coming from the substrate. It will be appreciated that the thermoelastic temperature change of the coating will always be in phase with the loading, whereas the phase of the 'main' effect will vary through the thickness and with frequency, so that the two may be in anti-phase or co-phase, or some intermediate condition, depending upon position, loading frequency, etc.

MacKenzie (1989) presents an analysis based upon such a model. The predictions, shown in figures 4.11 and 4.12, correlate well with the results in figures 4.9 and 4.10. In that work it was estimated that the thermoelastic response of the paint was some 11% of that of aluminium on an equal-strain basis.

This insight would explain the reports of meaningful results at kilohertz frequencies (see for example Turner and Pollard (1987)), whereas a simpler analysis, presuming an inert paint, would indicate total suppression of the signal. The reduced signal level will of course exacerbate noise problems.

The modelling by MacKenzie (1989) using the concepts of thermal waves (Carslaw and Jaeger 1959), reflection, interference, and absorption, has implications of practical importance. It constructs a picture of the paint emitting radiation at every point in its thickness, the radiation being partially re-absorbed on its way through, so that the total radiated flux at the surface is a summation of the energy emitted from different depths in the thickness thus allowing for re-absorption. It was only by using such a model that good agreement with the measured phase data could be obtained. This means that a minimum paint thickness is necessary to produce a fully developed radiant

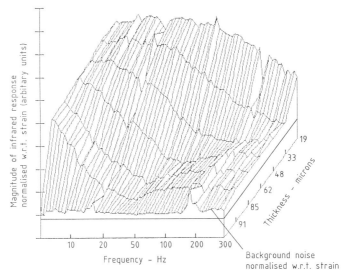

**Figure 4.9** Experimental results for the amplitude of the radiated signal from a plain specimen (versus coating thickness and loading frequency).

flux. Equally, considering the highly reflective nature of the paint–substrate interface, it is essential that there be adequate paint thickness to ensure that reflections of heat sources in the neighbourhood are adequately attenuated, for such reflections could be strongly coherent with the substrate's thermoelastic response due to synchronous specimen motion. It will be understood that even a fractional percentage reflection of a source that is some degrees kelvin different from the specimen could be of very significant magnitude in comparison to the thermoelastic temperature changes. MacKenzie (1989), on these considerations, recommends a paint thickness of 20 to 30 microns.

### 4.5.6 Correcting for the 'coating problem'

In principle it is clearly possible to correct for the variations in coating thickness. By measuring the paint thickness over the area, and knowing the thermal diffusivity of the paint and its thermoelastic properties, one could calculate the attenuation. Such an approach, however, does not seem practical. A more feasible approach is suggested by MacKenzie (1989), in which a scan of the response at a particular point is carried out over a range of excitation frequencies. By correlating the results with one of the theoretical models, an estimate can be made of a fictitious response at zero frequency; the frequency range employed must not include low frequencies such that internal condition or environmental effects could be significant.

The fictitious response should indicate the signal that would be obtained with a paint of perfect thermal diffusivity. Such a frequency scan could be incorporated as part of the routine operating procedure. The obvious penalty is the additional operating time involved. Figure 4.13 shows such a correlation against equation (4.14), the values of the unknown quantities being adjusted to give an optimal curve-fit. For a truer description, a combination of equations (4.14) and (4.19) ought to be utilised in the zero-frequency estimation process, but for the substrate in that particular test (aluminium) it was reckoned that the drag-down effect would be minimal.

### 4.5.7 Practical significance

On the basis of the analysis and the evidence available, it is clear that the effects of the paint coating may significantly influence the radiation fluctuations in both their amplitudes and their phases, and that the correlator output, from which the stress sum is determined, can be seriously affected by paint thickness variations unless the frequency is either quite low (5 Hz or less) or very high (200 Hz or greater). In the latter case a very thick coating is to be preferred, with the signal being then generated by the paint rather than the substrate.

In the frequency range 5–200 Hz, care should be taken to ensure that as far as possible the thickness is (i) adequate to ensure opacity, i.e., 20 microns or

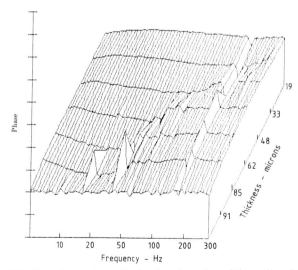

**Figure 4.10** Experimental results for the phase lag of the radiated signal from a plain specimen (versus coating thickness and loading frequency).

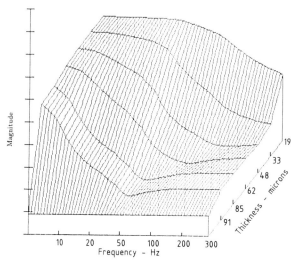

**Figure 4.11** Plot of the theoretical amplitude of the radiated signal using the MacKenzie interference model; corresponding to figure 4.9.

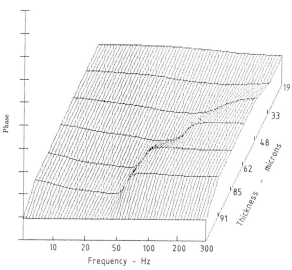

**Figure 4.12** Plot of the theoretical phase lag of the radiated signal using the MacKenzie interference model; corresponding to figure 4.10.

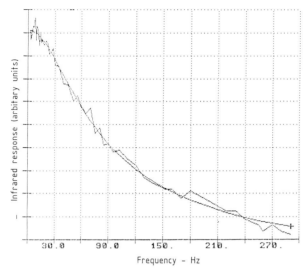

**Figure 4.13** Experimental signal values (versus loading frequency) and their correlation with equation (4.14).

greater; and (ii) as uniform as possible. (The minimum thickness criterion is relevant at all frequencies.) Until an established frequency-scanning routine is available, the measurement of the coating thickness is recommended, with an overcoating of any relatively thinly covered areas until the thickness is constant to within a few microns.

## 4.6 CONCLUDING REMARKS

The matter of non-adiabaticity is one of considerable complexity, giving grounds for sometimes conflicting criteria, which may be stated thus:

(i) Load-cycle frequency should be high so as to avoid effects due to internal heat leakage and environmental heat losses. The leakage is more relevant for the more conductive materials whereas the poorer conductors are more susceptible to environmental loss. When looking for the maximum detail, the frequency should be as high as possible.

(ii) On substrates whose emissivity is too low, or too non-uniform, to give satisfactory results without painting, the coating should be as uniform as possible.

(iii) Load-cycle frequency should be as low as possible in order to minimise effects due to point coatings—unless frequencies of hundreds of hertz can be used, when the paint's thermoelastic response controls the signal.

(iv) A paint thickness in the range of 20–30 microns should be aimed for unless the high-frequency range is feasible, when an even thicker coating is preferable.

(v) Phase information is valuable both as an indicator of the presence of the effects of thermal conduction, in substrates and in coatings, and in permitting more rigorous analyses of results.

The problems discussed in this chapter are under continuing investigation, and it is anticipated that correcting algorithms and procedures will become available in due course.

**ACKNOWLEDGMENTS**

Figures 4.1, 4.3 and 4.4 were published in 'Some Practical Limits to the Applicability of the Thermoelastic Effect' in *Experimental Stress Analysis*, H Wieringa (ed), pp 510–513, 1986, and are reprinted with permission of Kluwer Academic Publishers.

Figures 4.2–4.6 were published in 'Consideration of the Surface Response to Cyclic Thermoelastic Heat Generation' in *Stress Analysis by Thermoelastic Techniques*, SPIE vol 731, pp 44–53, 1987, and are reproduced by permission of SPIE.

# 5

# The Application of SPATE in Fracture Mechanics

## P Stanley

## 5.1 INTRODUCTION

The analysis and characterisation of crack-tip stress fields, and the determination of stress intensity factors and the relevant material parameters are topics of major importance in fracture mechanics. The stresses of interest are relatively localised and the stress gradients are high; in these circumstances the high sensitivity and resolution of the SPATE equipment and, in particular, the non-contact nature of the SPATE technique come together to offer a very attractive novel approach in crack-tip stress studies. This potential was recognised early in the development of the technique and the first confirmation of the feasibility of using the equipment for this purpose was given in a paper (Stanley and Chan 1985a) presented at the 1985 SEM Spring Conference on Experimental Mechanics held in Las Vegas, USA. Since then, successive quantitative studies of a range of specimens containing static cracks or machined slots under mode I or mode II loading have been described (Stanley and Chan 1986a,b, 1987a), and the Paris law crack-growth rate parameters for a propagating crack have also been determined (Stanley and Chan 1986a). A more recent study (Chan and Tubby 1988) has covered a particular case of mixed mode loading relating to a crack originating at a weld toe.

The work is described and discussed in this chapter.

## 5.2 THEORY

### 5.2.1 General

A crack in a flat plate is considered, under conditions of plane stress, subjected to mode I (opening) and mode II (sliding) loading (see figure 5.1). The Westergaard equations (Kassir and Sih 1975) give a solution for the elastic stresses $\sigma_x$, $\sigma_y$, and $\tau_{xy}$ (see figure 5.2) in the region of the crack tip in the form:

$$
\begin{bmatrix} \sigma_x \\ \\ \sigma_y \\ \\ \tau_{xy} \end{bmatrix} = \frac{K_\mathrm{I}}{\sqrt{2\pi r}}\cos(\theta/2) \begin{bmatrix} 1-\sin(\theta/2)\sin(3\theta/2) \\ \\ 1+\sin(\theta/2)\sin(3\theta/2) \\ \\ \sin(\theta/2)\cos(3\theta/2) \end{bmatrix}
$$

$$
+ \frac{K_\mathrm{II}}{\sqrt{2\pi r}} \begin{bmatrix} -\sin(\theta/2)(2+\cos(\theta/2)\cos(3\theta/2)) \\ \\ \sin(\theta/2)\cos(\theta/2)\cos(3\theta/2) \\ \\ \cos(\theta/2)(1-\sin(\theta/2)\sin(3\theta/2)) \end{bmatrix} \tag{5.1}
$$

where $K_\mathrm{I}$ and $K_\mathrm{II}$ are the mode I and mode II stress factors and $r$, $\theta$ are polar coordinates measured from the crack tip and the continuation of the line of the crack respectively (see figure 5.2).

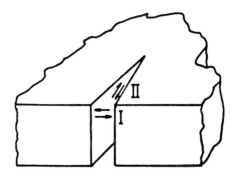

**Figure 5.1** A crack under mode I and mode II loading.

Equation (5.1) is the widely used form of a more general equation containing an infinite succession of higher-order terms; it is strictly valid only for an infinitely extensive specimen. The implications of using this form of the equation, in terms of specimen size and other practical details, have been discussed by Pukas (1987). In general, provided the coordinate point $(r, \theta)$ is not very close to the crack tip, nor affected by the proximity of the specimen edges, then equation (5.1) provides an adequate description of the stress state. In the immediate vicinity of the crack tip, the effects of local yielding invalidate the equation, and as the coordinate point approaches the specimen edges it becomes necessary to retain increasingly more terms from the general equation.

Under adiabatic cyclic loading, the SPATE signal ($S$) from a linear elastic solid is proportional to the change in the sum of the principal surface stresses $(\sigma_1 + \sigma_2)$ at the point under consideration (Stanley and Chan 1985b), i.e.

$$\sigma_1 + \sigma_2 = AS$$

where $A$ is a constant of proportionality or 'calibration constant'. (Details of the constant $A$ and the constituent terms in it are given by Stanley and Chan (1985b), who also describe direct methods for the determination of the constant. An expression for a related constant is given in the SPATE manual.)

The quantity $\sigma_1 + \sigma_2$ is the first invariant of a two-dimensional stress system and is therefore equal to the sum of the coordinate normal stresses $\sigma_x$ and $\sigma_y$; it follows that

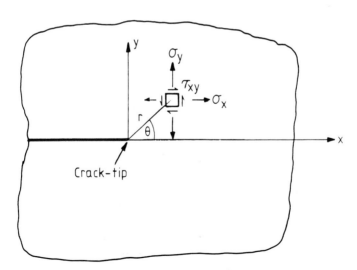

**Figure 5.2** Crack-tip coordinates.

$$\sigma_x + \sigma_y = AS. \tag{5.2}$$

Substituting from equation (5.1) for the stress terms in equation (5.2), the following important general relationship is obtained for the SPATE signal $S$ obtained from a point $(r, \theta)$ in a crack-tip stress field resulting from any combination of mode I and mode II loading

$$AS = \frac{2K_I}{\sqrt{2\pi r}} \cos(\theta/2) - \frac{2K_{II}}{\sqrt{2\pi r}} \sin(\theta/2). \tag{5.3}$$

Using equation (5.3) the stress intensity factors $K_I$ and $K_{II}$ can be determined from SPATE data for either single-mode or mixed-mode loading.

### 5.2.2 Mode I loading

The case of pure mode I loading is considered first. It is envisaged that the SPATE data will be appraised initially by means of a series of signal line plots taken parallel to the line of the crack (i.e. $y=$ constant, see figure 5.2). Equation (5.3) is developed by putting $K_{II}=0$ and by replacing $r$ in the $K_I$ term by $y/\sin\theta$ to give

$$AS = \frac{2K_I}{\sqrt{2\pi y}} \sqrt{\sin\theta} \cos(\theta/2). \tag{5.4}$$

It is readily shown from equation (5.4) that the partial derivative of $S$ with respect to $\theta$ (i.e. $\partial S/\partial\theta$) is zero when $\theta = 60°$, and it follows that the maximum signal $(S_{max})$ in a line plot taken parallel to the crack occurs at $\theta = 60°$. It also follows from equation (5.4) by substituting this value of $\theta$ that

$$S_{max} = \frac{2K_I}{A\sqrt{2\pi y}} \left(\frac{\sqrt{3}}{2}\right)^{1/2} \frac{\sqrt{3}}{2} \tag{5.5}$$

or, alternatively

$$y = \left(\frac{3\sqrt{3}\,K_I^2}{4\pi A^2}\right) \frac{1}{S_{max}^2}. \tag{5.6}$$

It is evident from equation (5.6) that $y$ and $1/S_{max}^2$ are linearly related and that, provided the constant $A$ is known, $K_I$ can be obtained directly from the gradient of a graph of $y$ versus $1/S_{max}^2$. The choice of equation (5.6) for this purpose rather than equation (5.5) (i.e. $S_{max}$ plotted against $1/\sqrt{y}$) is important since, with the former, the linearity of the relationship is retained with an arbitrary systematic error in $y$ and the exact position of the crack need not therefore be known. This is not so with equation (5.5). The graphical approach has some advantages in this context but clearly a simple numerical least-squares fit approach can be used for the determination of the 'gradient'.

Alternatively, the analysis for mode I loading can be based on signal plots along lines perpendicular to the crack and ahead of the crack tip (i.e. $x=$constant, see figure 5.2). For such plots it follows by symmetry that the maximum signal $S_{max}$ occurs for $y=0$ (i.e. $\theta=0$). With $K_{II}=0$, $\theta=0$ and $r=x$, equation (5.3) becomes

$$AS_{max} = \frac{2K_I}{\sqrt{2\pi x}} \tag{5.7}$$

or

$$x = \left(\frac{2K_I^2}{\pi A^2}\right)\frac{1}{S_{max}^2}. \tag{5.8}$$

$K_I$ can be derived therefore from the gradient of the linear graph of $x$ versus $1/S_{max}^2$. Again, the particular form of equation (5.8) is important since linearity is ensured with a systematic error in $x$, i.e. an exact knowledge of the position of the crack tip is not required and $x$ can be measured from an arbitrary datum. A further feature of the graphical approaches represented in equations (5.6) and (5.8) is that regions of the stress field over which the simple Westergaard equations (equation 5.1) do not apply are readily identified as regions where the linearity of the $1/S_{max}^2$ plot breaks down.

Equation (5.8) also represents the variation of the signal with $x$ along the line $y=0$ ahead of the crack, and consequently a signal line scan along this line is sufficient for the determination of $K_I$; it is obtained as before from the gradient of a graph of $x$ versus $1/S^2$. This method is particularly relevant for the case of a propagating crack (see later).

### 5.2.3 Mode II loading

For the case of mode II loading (i.e. $K_I=0$), again with the substitution $r=y/\sin\theta$ equation (5.3) becomes

$$AS = -\frac{2K_{II}}{\sqrt{2\pi y}}\sqrt{\sin\theta}\,\sin(\theta/2). \tag{5.9}$$

The partial derivative of $S$ with respect to $\theta$ is determined and equated to zero to give the $\theta$-value of the maximum (or minimum) signal along a line plot parallel to the crack length (i.e. $y=$constant). The derived $\theta$-values are $\pm120°$ for positive and negative $y$-values respectively. Taking the positive $\theta$-value, the following relationship between the maximum (negative) signal $S_{max}$ along a line of constant positive $y$ and the $y$-coordinate of the line is obtained:

$$y = \left(\frac{3\sqrt{3}K_{II}^2}{4\pi A^2}\right)\frac{1}{S_{max}^2}. \tag{5.10}$$

For $\theta = -120°$ (i.e. along a line of constant negative $y$) the peak signal is positive and a similar equation is obtained with a minus sign on the right-hand side.

Equation (5.10) is similar to equation (5.6), and the procedure described after the latter equation for the determination of $K_{\mathrm{I}}$ applies equally for the determination of $K_{\mathrm{II}}$ from the signal maxima, $S_{\max}$, observed in a series of line scans parallel to a crack under mode II loading.

$K_{\mathrm{II}}$ can also be determined from signal line plots perpendicular to the crack length and ahead of the crack tip (i.e. $x =$ constant, see figure 5.2). These plots exhibit numerically equal minimum and maximum signals at $\theta = 60°$ and $-60°$ respectively and a relationship between the signal peaks and the positive $x$-coordinate is readily obtained. These plots, however, are less suitable for the determination of $K_{\mathrm{II}}$ than the constant-$y$ plots (equation (5.10)) and this alternative approach has not been used.

### 5.2.4 Mixed-mode loading

For mixed-mode loading in which both $K_{\mathrm{I}}$ and $K_{\mathrm{II}}$ are finite, equation (5.3) is required without simplification. Several approaches are possible. The factors can be found from a routine data-fitting operation using discrete signal values obtained at a number of coordinate points, and the effects of random errors can be considerably reduced when a sufficient number of individual signal readings are used. However, significant errors may occur in the values of $K_{\mathrm{I}}$ and $K_{\mathrm{II}}$ obtained in this way because the $r, \theta$ coordinate values will be inaccurate due to uncertainty in the exact position of the crack tip and, moreover, it cannot be ensured that the relevant data points lie within the range of validity of equation (5.3). Alternatively, $K_{\mathrm{I}}$ and $K_{\mathrm{II}}$ can be derived from signal values along the $\theta = 0$ line, for which the $K_{\mathrm{II}}$ term in equation (5.3) is zero; knowing $K_{\mathrm{I}}$, $K_{\mathrm{II}}$ can then be obtained from readings along another constant-$\theta$ line (e.g. $\theta = \pm 90°$).

Chan and Tubby (1988) based their mixed-mode loading analysis on the results of a series of signal plots along straight lines inclined at an angle $\beta$ to the line of the crack (see figure 5.3), and on the determination of the $\theta$ coordinate ($\theta_{\max}$) of the maximum signal ($S_{\max}$) along these lines. These authors worked in terms of the 'geometric correction factors' $Y_{\mathrm{I}}$ and $Y_{\mathrm{II}}$ defined as

$$Y_{\mathrm{I}} = \Delta K_{\mathrm{I}} / \Delta \sigma \sqrt{\pi a}$$

$$Y_{\mathrm{II}} = \Delta K_{\mathrm{II}} / \Delta \sigma \sqrt{\pi a}$$

(5.11)

where $\Delta \sigma$ and $a$ are the range of the 'remote applied stress' and the crack length, respectively. (A note on the use of $\Delta K_{\mathrm{I}}$ and $\Delta K_{\mathrm{II}}$ in equations (5.11) is given at the end of this section.) With these substitutions and with $r$ replaced by $y'/\sin(\beta+\theta_{max})$ (see figure 5.3) it follows from equation (5.3) that

$$y' = \frac{2a\sin(\beta+\theta_{max})\Delta\sigma^2}{A^2 S^2_{max}}\left[Y_I\cos\left(\frac{\theta_{max}}{2}\right) - Y_{II}\sin\left(\frac{\theta_{max}}{2}\right)\right]^2 \qquad (5.12)$$

and therefore that

$$Y_I\cos(\theta_{max}/2) - Y_{II}\sin(\theta_{max}/2) = \left(\frac{A^2 m^+}{2a\sin(\beta+\theta_{max})\Delta\sigma^2}\right)^{1/2} \qquad (5.13)$$

where $m^+$ is the gradient of the linear plot of $y'$ versus $1/S^2_{max}$ for positive values of $y'$.

A second equation in $Y_I$ and $Y_{II}$ is obtained for negative $y'$-values (see figure 5.3) of the form

$$Y_I\cos(-\theta_{max}/2) - Y_{II}\sin(-\theta_{max}/2) = \left(\frac{A^2 m^-}{2a\sin(\beta-\theta_{max})\Delta\sigma^2}\right)^{1/2} \qquad (5.14)$$

where $m^-$ is the gradient of the linear plot of $y'$ versus $1/S^2_{max}$ with negative $y'$-values.

Using the appropriate values of $A$, $a$, $\beta$ and $\Delta\sigma$, and having obtained $\theta_{max}$ from the experimental results, Chan and Tubby (1988) derived values of $Y_I$ and $Y_{II}$ by simultaneous solution of equations (5.13) and (5.14).

(Note: Up to equations (5.11) the stresses and the stress intensity factors $K_I$ and $K_{II}$ are used without the prefix $\Delta$. In the present context, however, the '$\Delta$' prefix is particularly relevant since the SPATE signal is necessarily associated with a stress change. For this reason the $\Delta$ prefix is used in equations (5.11) and subsequently where appropriate. The quantities $Y_I$ and $Y_{II}$ and $k_I$ and $k_{II}$ (see later) are ratios and for them the $\Delta$ prefix is inappropriate.)

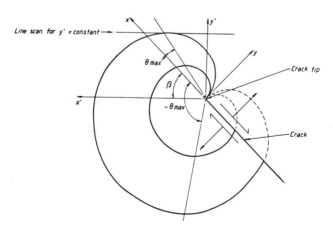

**Figure 5.3** Coordinates for mixed-mode loading.

### 5.2.5 Crack propagation under mode I loading

Except for the initial and final stages of growth (Paris and Erdogan 1963), the increase in the length ($a$) of a propagating fatigue crack per cycle ($N$) of applied load is related to the stress intensity range over the load cycle ($\Delta K_1$) by the Paris law (Paris and Erdogan 1963) expression

$$\mathrm{d}a/\mathrm{d}N = C(\Delta K_1)^m \tag{5.15}$$

where the coefficient $C$ and the exponent $m$ are characterising parameters for the particular material (Smith 1979). In demonstrating the validity of this expression, and deriving values for $C$ and $m$, it is necessary to obtain simultaneous values of $\Delta K_1$ and $\mathrm{d}a/\mathrm{d}N$. Both of these quantities can be determined from signal scans along the line of the crack (i.e. $y = 0$, see figure 5.2). The determination of $\Delta K_1$ is based upon equation (5.8) and has been outlined at the end of the earlier section on mode I loading. The value of $\mathrm{d}a/\mathrm{d}N$ is obtained as the gradient of a plot of the crack length versus the number of cycles ($N$). In the work to be described, a central crack in a wide plate was studied and the double crack length ($2a$), after a certain number of cycles, was measured as the distance between the two signal maxima in a single line scan covering both crack tips. The line scan could be completed in a relatively small number of loading cycles and it was satisfactory to take as $N$ the average $N$-value pertaining to the beginning and end of the scan. Minor difficulties may arise in ensuring that the crack tips are accurately positioned on the scan line. In practice these difficulties can be avoided by taking vernier microscope readings of crack length or by use of an alternative approach based on an analysis of the full signal scan over the tip region of the propagating crack, as described below.

The SPATE signal display is generated in a series of horizontal lines from the top to the bottom of the scanned 'frame'. In the case of a propagating crack, each successive line of the complete scan is displaced in the direction of crack propagation by an amount equal to the distance through which the crack tip has advanced during the time interval between successive line scans. The effect of this is to produce a 'skewing' or 'shearing' of the signal display. This is illustrated in figures 5.4(*a*) and 5.4(*b*) which show, respectively, typical signal contours for a stationary crack under mode I loading, generated numerically from equation (5.3), and signal contours for a crack propagating under mode I loading with $\mathrm{d}a/\mathrm{d}N$ equal to $1 \times 10^{-3}$ mm per cycle.

In analysing the skewed signal display, the upper portions of two adjacent contours of $\sigma_x + \sigma_y$ (see equation (5.2)) produced in the scanning of a propagating crack are considered (see figure 5.5) The uppermost points of the outer contour, A, is generated when the crack tip is at B; the uppermost point, C, on the inner contour is generated when the crack tip has advanced to D, a distance $\Delta a$ beyond B. The line AC makes an angle $\varphi$ with a horizontal line, parallel to the crack and the $x$-axis, through C. Had the crack been stationary, the uppermost point of the inner contour (shown dotted in the figure) would have been at E and the angle $\varphi$ would have been 60°. Also shown in figure 5.5 are $\Delta x_m$ and $\Delta x_s$, the differences between the $x$-coordinate of point A and those of points C and E, respectively. From the figure it can be seen that

$$\Delta a = \Delta x_s - \Delta x_m = \Delta y \left( \frac{1}{\tan 60°} - \frac{1}{\tan \varphi} \right) \tag{5.16}$$

where $\Delta y$ is the difference in the $y$-coordinates of points A and C.

If $Y$ is the full vertical dimension of the completed scan and $T$ is the total scan time, then the time interval between generating the signal at A and the signal at C (i.e. the time for the crack to advance from B to D) is

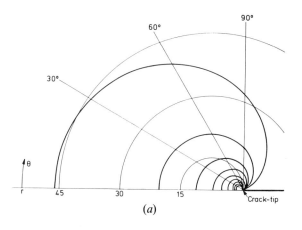

(a)

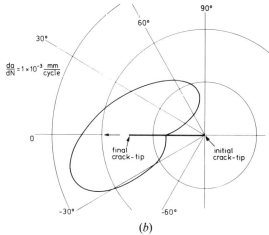

(b)

**Figure 5.4** Typical signal contours for (a) a static crack and (b) a propagating crack.

approximately $T\Delta y/Y$. The number of load cycles ($\Delta N$) applied during this time interval is given by

$$\Delta N = f\frac{T\Delta y}{Y} \tag{5.17}$$

where $f$ is the cyclic load frequency.

From equations (5.16) and (5.17) it follows that

$$\frac{\Delta a}{\Delta N} \approx \frac{da}{dN} = \frac{Y}{fT}\left(\frac{1}{\sqrt{3}} - \cot\varphi\right). \tag{5.18}$$

In the limit, $\varphi$ is the slope of the locus of the signal contour peak and the approximation in equation (5.17) becomes insignificant. Therefore, using equation (5.18), $da/dN$ may be determined from the scan parameters and from the signal contours obtained from the vicinity of the propagating crack tip.

If a series of corresponding values of $\Delta K_1$ and $da/dN$ for a propagating crack are obtained as described above, the Paris law parameters $C$ and $m$ (see equation (5.15)) are readily derived from the intercept and gradient of a log–log plot of this data.

## 5.3 EXPERIMENTAL WORK

### 5.3.1 General

The procedures used in this series of SPATE applications were standard in all respects. The test material was mild steel throughout. The test surface of each

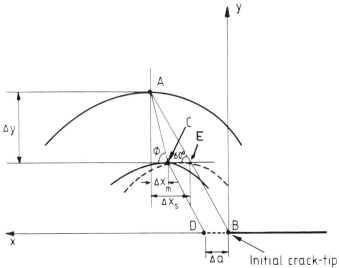

**Figure 5.5** Adjacent signal contours for a propagating crack.

specimen was cleaned with acetone and coated immediately prior to testing with two 'passes' of RS matt black aerosol paint. (It has been shown (Stanley and Chan 1985a) that this surface treatment does not impair the SPATE signal.) In defining and locating on the specimen the area to be scanned it was found convenient to attach temporarily to the test surface a small strip of transparent graph paper; this was removed in the test proper. The SPATE equipment used in the mode I and mode II work, and for the propagating crack investigation, was the 'prototype' equipment used by Stanley and Chan (1987b) in their work at the University of Manchester. A trivial feature of this equipment was that the vertical scale of the SPATE scan display was reduced by a factor of 0.67 relative to the horizontal scale; as a consequence of this the SPATE scans shown later in the chapter for these investigations are distorted to this degree in terms of the horizontal and vertical scales. A SPATE 8000 system was used for the mixed-mode loading work, which was carried out at the Welding Institute. Several loading machines were used. In each test the reference signal for the SPATE correlator was taken from the load transducer in the machine.

### 5.3.2 Mode I loading

The specimen used for the mode I loading study was a flat rectangular steel plate 300 mm wide, 350 mm long and 1.6 mm thick, with a 3 mm diameter central circular hole, as shown in figure 5.6. Two steel reinforcing strips were welded across the width of the plate at each end; the specimen was secured into the shackles of the test machine by means of seven equi-spaced 10 mm diameter pins at each end so as to produce a uniformly distributed tensile loading across the central part of the plate. The machine used was a 250 kN servo-controlled hydraulic fatigue testing machine operating at a cyclic frequency of 10 Hz. (This was well in excess of the minimum frequency of approximately 2 Hz required for adiabatic behaviour in this material.)

Prior to the SPATE testing, the specimen was subjected to cyclic tensile loading so that fatigue cracks grew symmetrically across the plate from the peak stress positions on the hole boundary until they reached a required length. The amplitude of the applied cyclic load was then reduced so that the cracks did not propagate and the SPATE scanning over a selected rectangular frame around one crack tip was commenced. In order to obtain maximum resolution over the crack-tip region the detector was brought to within 205 mm of the specimen surface; the resulting scan resolution was $0.6 \times 0.6$ mm. Having completed a SPATE test, the cracks were extended to a new stable length and a further test was carried out. Results were obtained for four crack lengths.

A typical SPATE signal display obtained from the tip region of a non-propagating crack subjected to mode I loading is shown in figure 5.7; the crack length in this case was approximately 17 mm, giving a crack length to semi-plate-width ratio ($a/b$) of 0.113. The symmetry of the display is noteworthy (the line of the crack coincides with the axis of symmetry, and the crack tip coincides with the peak signal) and, in spite of the linear distortion inherent in signal displays from the prototype equipment, the close general correspondence between the experimental results and the numerically derived signal contours illustrated in figure 5.4($a$) is evident.

A typical signal plot taken from the stored scan data along a line parallel to the line of the crack (in this case $y = -1.5$ mm, see figure 5.2) is shown in figure 5.8. (The axis graduations are arbitrary; the vertical axis shows signal strength, the horizontal axis shows the position along the line of the scan.) The signal maximum occurring at $\theta = 60°$ (see equations (5.4) and (5.5)) is clearly defined.

Figure 5.9 has been constructed from a series of such plots taken both above and below the crack; it is a portrayal of equation (5.6) and shows the $y$-coordinate of the line plotted against the reciprocal of the square of the maximum signal ($1/S^2_{max}$) (the origin of the $y$-coordinates has been taken arbitrarily at a point approximately 10 mm above the line of the crack). The

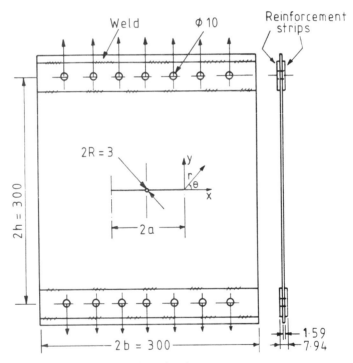

Material: Black mild steel
Hole dia./plate width   R/b = 0·01
Test section height/plate width: h/b = 1
Dimensions in mm

**Figure 5.6** Specimen for mode I loading.

two best-fit straight lines through the results are shown in the figure and the degree of the fit about the lines is such as to clearly demonstrate the validity of equation (5.6). The lack of fit near the crack tip is to be associated with localised non-elastic behaviour which appears to extend about 0.5 mm on either side of the crack.

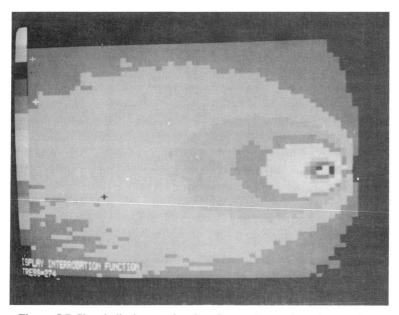

**Figure 5.7** Signal display at the tip of a static crack under mode I loading.

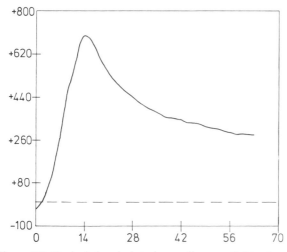

**Figure 5.8** Typical signal plot along a line parallel to the crack.

The calibration constant $A$ (see equations (5.2) and (5.6)) is dependent upon the test material and on factors specific to the particular equipment. A value of $A$ of 0.393 N mm$^{-2}$ per unit SPATE signal had been established for steel using the prototype equipment (Stanley and Chan 1985b). With this value of $A$, a $\Delta K_I$-value for the crack under investigation was readily obtained from the gradient of the best-fit lines shown in figure 5.9 (see equation (5.6)). It is convenient to present such values in non-dimensional form as $k_I = \Delta K_I / \Delta\sigma\sqrt{\pi a}$. Values of $k_I$ for several different crack lengths are given in table 5.1, together with corresponding numerical values obtained using the boundary collocation method (Rooke and Cartwright 1976).

Figure 5.10 shows a sample from a series of signal line plots taken perpendicular to the line of the crack ($x = 3.8$ mm, see figure 5.2); the symmetry is evident. Figure 5.11 is a graph of $x$ versus $1/S_{max}^2$ taken from this series. A clear linear range (3 mm $< x <$ 11 mm) is evident, conforming with equation (5.8). Again there is a significant trend away from linearity as the crack tip is approached and, additionally in this case, as the plate edge is approached; this latter trend is an indication that higher-order terms in equation (5.1) are required in this region. Again a value of $k_I$, the non-dimensional version of $K_I$, has been derived from the gradient of the best-fit line through the figure 5.11 data and from similar plots for a number of increasing crack lengths. The results are included in table 5.1.

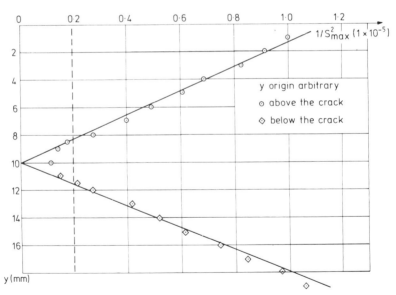

**Figure 5.9** A graph of $y$ versus $1/S_{max}^2$ for mode I loading.

The general level of agreement between the two independent sets of experimental results and the numerical results in table 5.1 is good. Experimental values derived from line scans parallel to the crack line (i.e. along lines of constant $y$) tend to be marginally smaller than other comparable values.

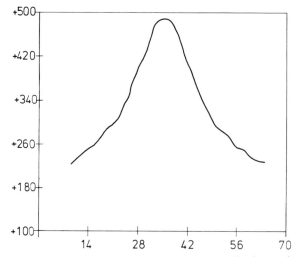

**Figure 5.10** Typical signal plot along a line perpendicular to the crack.

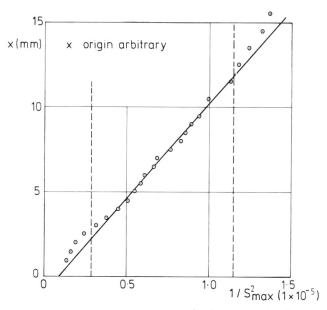

**Figure 5.11** A graph of $x$ versus $1/S_{max}^2$ for mode I loading.

### 5.3.3 Mode II loading

Several forms of specimen have been used for mode II loading (Stanley and Chan 1986b, 1987a). The 'cracks' in each case were simulated by machined slots 0.8 mm wide and with 'blunt' ends; no attempt was made to simulate a real crack tip at the notch root. The test machine and test details were the same as those described above for mode I loading.

The first specimen (designated the 'pull–pull' specimen) is illustrated in figure 5.12. The design was developed from a compact shear specimen studied both numerically and photoelastically by Chisholm and Jones (1977). The specimen seemed attractive and the test work was straightforward, but there were large systematic differences (Stanley and Chan 1986b) between the results derived from the thermoelastic data and those obtained from Chisholm and Jones's work (1977). At the time of writing these differences have not been satisfactorily resolved and it is clear that further work will be required in this particular area.

Bank-Sills *et al* (1984) have described and analysed a 'mode II fracture specimen' and this has been assessed by means of the SPATE technique. The specimen is shown in figure 5.13. The loading, which may be tensile (i.e. machine jaws opening) or compressive (i.e. machine jaws closing), is applied in the central plane of the specimen at the uppermost and lowest points. The central vertical plane of the central portion of the specimen is in a state of pure shear and a crack or slot in that plane will be subjected to mode II loading. Results are presented here for two such slotted specimens: (i) one with a machined edge slot 20 mm deep and loaded in compression, and (ii) one with a spark-eroded central slot 18 mm long, 0.8 mm wide, loaded in tension. Mid-plane loading was ensured by applying the load through balls or 'ball-and-stirrup' connections, the balls resting in 3 mm diameter holes accurately positioned so that the axes coincided with the mid-plane of the specimen. The occurrence of significant out-of-plane distortions observed

**Table 5.1** Values of $k_I$.

|  | Experimental values | | Numerical values (Rooke and Cartwright 1976) |
|---|---|---|---|
| $a/b$ | Along line of constant $y$ | Along line of constant $x$ | |
| 0.113 | — | 1.06 | 1.05 |
| 0.166 | 1.04 | 1.10 | 1.07 |
| 0.210 | 1.01 | 1.10 | 1.08 |
| 0.305 | 1.04 | 1.07 | 1.12 |

with specimens 1.6 mm thick was avoided when the thickness was increased to 12 mm.

Figure 5.14 shows typical signal plots along lines parallel to the edge slot on each side of the slot. Signal maxima were read from a series of such plots. Figure 5.15 is the resulting graph of $y$ versus $1/S^2_{max}$ (the $y$ origin was at a convenient point to the left of the scan). The corresponding graph for the specimen with a central slot is shown in figure 5.16. Both parts of the graphs in figures 5.15 and 5.16 show well-defined linear portions with marked non-linear trends close to the slot root. $\Delta K_{II}$-values were derived from the gradients of the best-fit straight lines through the linear portions of these plots using equation (5.10). In order to allow direct comparison with the results obtained from photoelastic (Arcan and Bank-Sills 1982) and FE (Bank-Sills *et al* 1984) studies on similar specimens, non-dimensional stress intensity factors $(k_{II})$ were derived from the $\Delta K_{II}$-values by dividing by the quantity $\Delta \tau \sqrt{\pi a}$, where $\Delta \tau$ is the average shear stress amplitude over the unslotted central section of the specimen and $a$ is the length of the edge slot and the semi-length of the central slot. Values of $k_{II}$ from the thermoelastic, photoelastic and FE work are given in table 5.2, together with the dimensional

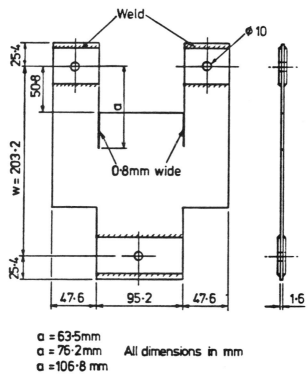

a = 63·5 mm
a = 76·2 mm     All dimensions in mm
a = 106·8 mm

**Figure 5.12** The 'pull–pull' specimen.

ratios $a/c$ and $b/c$ ($b$ and $c$, semi-width and semi-depth, respectively, of the central section of the specimen). The small differences in these ratios for otherwise comparable specimens in the table should be noted. It is also relevant that no information on the slot-width and slot-end details for the photoelastic and numerical work is available.

The gradients of the linear portions of the graphs in figures 5.15 and 5.16 are well defined and possible random errors in the derived $k_{II}$-values are considered to be small. Systematic errors may occur as a result of the use of the previously determined $A$-value (see equation (5.2)) for steel. The coefficient of variation in this quantity was 6.6% (Stanley and Chan 1985b) and it must be assumed that errors of this order may be present in the tabulated $k_{II}$-values. For the edge-slot specimen (table 5.2) there is good agreement between the two thermoelastic results themselves and between these and the photoelastic and numerical results. The marginal increase in the thermoelastic results relative to the other two may be associated with the slightly greater $a/c$ and $b/c$ values. The $k_{II}$-values obtained for the central-slot specimen showed a degree of cross-symmetry, and the mean value of 0.91 was considerably smaller than the independent photoelastic and FE values, which were themselves comparable. There is no ready explanation for this in terms of the differences in the dimensional ratios or other minor differences

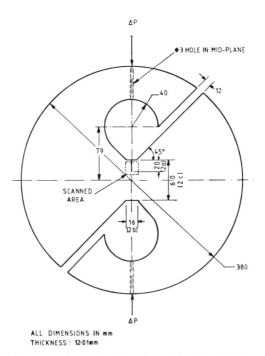

ALL DIMENSIONS IN mm
THICKNESS : 12·01mm

**Figure 5.13** Specimen for mode II loading (after Bank-Sills *et al* 1984).

between the specimens. The relevant area of the specimen was relatively small and the stress gradients were comparatively high but clearly further work is essential for a full understanding of these results.

The third mode II specimen was similar in essentials to the previous specimen but much simpler geometrically and, therefore, much easier to manufacture; it is illustrated in figure 5.17. (It had been demonstrated earlier (Stanley and Chan 1985b) that the central section of this specimen, loaded as indicated, experiences a pure-shear stress distribution.) One specimen with an edge-slot 27.5 mm long (see figure 5.17) was tested. The experimental and procedural details were the same as those for previous tests. Figure 5.18

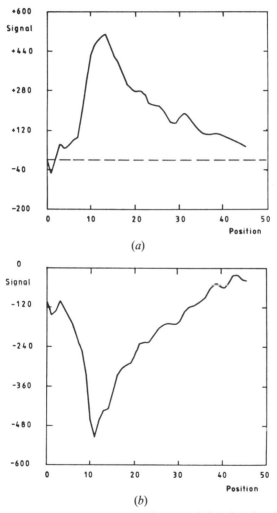

(a)

(b)

**Figure 5.14** Typical signal plots along lines parallel to the edge slot: (a) to the left, and (b) to the right.

shows the $y$ versus $1/S^2_{max}$ graph obtained from the full set of signal plots along lines parallel to the slot (the $y$-coordinate origin was arbitrary). The $\Delta K_{II}$-values derived from the gradients of the best-fit lines through the linear parts of the figure 5.18 graph differed by about 1%, and the mean $k_{II}$-value ($k_{II} = \Delta K_{II}/\Delta\tau\sqrt{\pi a}$) was 1.36. For this configuration Richard (1980) gives a $k_{II}$-value of 1.49, but this figure cannot be confirmed from the unreferenced formula given in the paper.

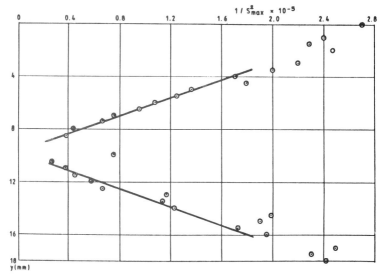

**Figure 5.15** A graph of $y$ versus $1/S^2_{max}$ for the edge slot.

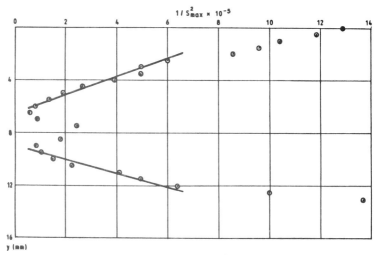

**Figure 5.16** A graph of $y$ versus $1/S^2_{max}$ for the central slot.

**Table 5.2** $k_{II}$-values from a shear specimen

| Thermoelastic result | | Photoelastic work (Arcan and Bank-Sills 1982) | FE work (Bank-Sills *et al* 1984) |
|---|---|---|---|
| *(a)* Specimen with edge slot | | | |
| $a/c$ | 0.667 | 0.637 | 0.637 |
| $b/c$ | 0.267 | 0.200 | 0.200 |
| $k_{II}$ | $\begin{cases} 1.040^a \\ \\ 1.058^b \end{cases}$ | 0.984 | 1.035 |
| *(b)* Specimen with central slot | | | |
| $a/c$ | 0.300 | 0.333 | 0.300 |
| $b/c$ | 0.267 | 0.267 | 0.250 |
| $k_{II}$ | $\begin{cases} 0.914^c \quad 0.876^e \\ \\ 0.831^d \quad 1.006^f \end{cases}$ | 1.222 | 1.195 |

Note:  [a] right-hand side of slot      [d] left-hand side, upper end of slot
         [b] left-hand side of slot        [e] left-hand side, lower end of slot
         [c] right-hand side, upper end of slot    [f] left-hand side, lower end of slot.

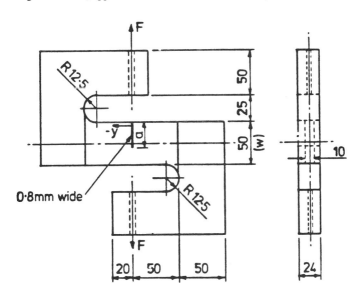

$a = 27.5$mm     All dimensions in mm

**Figure 5.17** The 'pure-shear' specimen.

### 5.3.4 Mixed-mode loading

The objective of the mixed-mode loading investigation (Chan and Tubby 1988) was to determine both stress-intensity factors at a weld-toe crack in a transversely loaded fillet-welded joint and to compare these factors with values obtained from a two-dimensional FE analysis. The test specimen (see figure 5.19) was 1780 mm in overall length with a uniform thickness of 12.75 mm. The fillet angle was 45°. A through-thickness starter slot, 5 mm deep and 0.15 mm wide, was cut normal to the specimen axis at the fillet toe using a slitting wheel. Strain gauges were attached to the specimen, as indicated in figure 5.19, for reference in the determination of the calibration factor ($A$). A 1000 kN hydraulic testing machine, operating at 6.6 Hz, was used to cyclically load the specimen. A fatigue crack was initiated at the slot-root and was allowed to propagate a uniform total depth (i.e. including the 5 mm depth of the slot) of approximately 8.5 mm. (It is noteworthy that, whilst very small in relation to the specimen width, this full depth is significantly smaller than the specimen thickness.) The crack growth occurred in a direction inclined 12° below the starter slot, as shown in the inset in figure 5.19. At this point the load amplitude was reduced by 50% to prevent further crack propagation and, after the usual preliminaries, a SPATE scan was taken over the region around and ahead of the crack tip. The SPATE detector was positioned 266 mm from the specimen, giving a spatial resolution slightly greater than 0.5 mm × 0.5 mm. After the SPATE scan had been

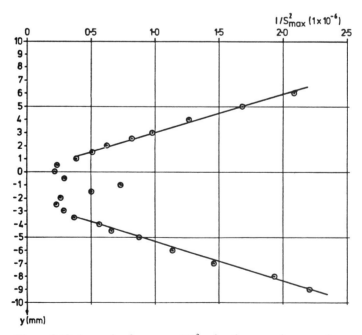

**Figure 5.18** A graph of $y$ versus $1/S_{max}^2$ for the pure-shear specimen.

completed, the load amplitude was increased and the crack was allowed to propagate a further 1 mm. The load amplitude was then reduced again and a second SPATE scan completed.

The data analysis followed the pattern outlined in the theory section (equations (5.11)–(5.14)). The maximum signal values ($S_{max}$) along each of a

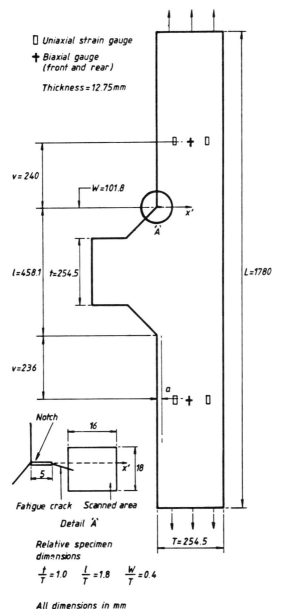

**Figure 5.19** Specimen for mixed-mode loading.

series of horizontal lines above and below the crack (i.e. along lines of constant $y'$) were obtained and plotted in the form of $y'$ versus $1/S^2{}_{max}$. The results are shown in figure 5.20 (the $y'$-coordinate origin is arbitrary). The gradients of the best-fit lines through the results above and below the crack were determined for eventual use in equations (5.13) and (5.14) as $m^+$ and $m^-$ respectively. The $\theta_{max}$-values for positive and negative values of $y'$ were obtained directly from a graph of the $x'$- and $y'$-coordinates of the $S_{max}$ positions. The calibration constant, $A$, was obtained as the ratio of the remote applied stress range to the signal level along a transverse line of the specimen adjacent to the strain gauges (see figure 5.19).

Having evaluated these relevant quantities, equations (5.13) and (5.14) were solved simultaneously for $Y_I$ and $Y_{II}$. Values obtained for the two crack lengths studied are given below in table 5.3.

The investigation also included a FE analysis of the shorter weld-toe crack in which a special crack-tip element was used to model the singularity at the crack tip and two loading conditions were studied; namely, (i) a uniform axial tensile stress, and (ii) a uniform end displacement. Values of $Y_I$ and $Y_{II}$ were derived from the FE results in three ways: (i) from the displacement

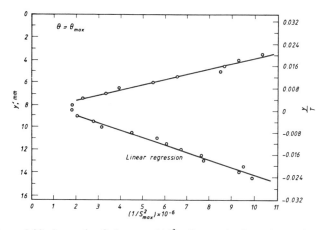

**Figure 5.20** A graph of $y'$ versus $1/S^2_{max}$ for a mixed-mode specimen.

**Table 5.3** $Y_I$ and $Y_{II}$ values (weld-toe cracks).

| $a$ (mm) | $a/T$[a] | $Y_I$ | $Y_{II}$ |
|---|---|---|---|
| 8.5 | 0.033 | 1.33 | 0.11 |
| 9.5 | 0.037 | 1.10 | 0.17 |

[a] $T$ = plate width.

components, (ii) from the stress sum along the line $\theta=0$, and (iii) from the stress sum along the line $\theta=\pm 90°$. These values are included in table 5.4.

It can be seen from table 5.4 that the numerical results consistently over-estimate the experimental results. For $Y_I$, the excess increased from a few per cent for the values derived from the stress sum along the line $\theta=\pm 90°$ to 20% for the values based on displacement values. For $Y_{II}$, the numerical results were consistently about 100% greater than the experimental results. Possible reasons for these discrepancies are discussed by Chan and Tubby (1988) and it is accepted that further work is necessary relating to both the inconsistencies in the numerical results themselves and the differences between the numerical and experimental results. It is the present writer's view that a great deal could be gained by extending this work.

### 5.3.5 Crack propagation under mode I loading

The specimen used in the crack propagation study was much the same as that shown in figure 5.6, but the thickness was 6 mm instead of 1.6 mm. The loading was by means of a 400 kN Dartec fatigue machine operating at a cyclic frequency of approximately 3.6 Hz. Mode I fatigue cracks were propagated from the roots of two fine saw-cuts made at a small circular hole at the centre of the plate. The SPATE detector was again positioned 205 mm from the specimen surface and the usual preparatory treatment was applied. After the effect of the initial saw-cuts had become negligible, line scans along the line of the cracks (i.e. $y=0$, see figure 5.2) and full scans over the region of

**Table 5.4** Numerical and experimental results ($a/T=0.033$).

| Geometric correction factor | Loading condition | Numerical values based on | | | Experimental values |
|---|---|---|---|---|---|
| | | Displacement extrapolation | Stress sum for $\theta=0$ | Stress sum for $\theta=\pm 90°$ | |
| $Y_I$ | Uniform end displacement | 1.62 | 1.53 | 1.38 | 1.33 |
| | Uniform axial stress | 1.57 | 1.49 | 1.34 | |
| $Y_{II}$ | Uniform end displacement | 0.21 | — | 0.20 | 0.11 |
| | Uniform axial stress | 0.20 | — | 0.19 | |

one of the crack tips were taken at convenient total load cycle numbers. Additionally, microscope readings of the crack length on both surfaces of the specimen were taken at regular intervals. (The front surface was cleaned with an acetone swab for these optical measurements.)

A typical scan along the line of the cracks and across the full width of the plate is shown in figure 5.21; the signal peaks at the crack tips are clearly evident. The total (i.e. double) crack length was obtained directly from this line scan by scaling. Values derived in this way are plotted against the average cycle number, $N$, for the scan in figure 5.22, together with results obtained from the microscope measurements. The small effects of the finite resolution of the SPATE equipment are evident and it is clear that the crack fronts were not quite normal to the plane of the plate. Values of $da/dN$ obtained from the gradient of the SPATE curve and the front-face microscope curve in figure 5.22 are plotted against $N$ in figure 5.23; the differences between the two sets of results are insignificant. Three values of $da/dN$ derived from the full SPATE signal display as outlined in the theory section (see equation (5.18)) are also shown in figure 5.23; they conform reasonably with the other results.

Finally, a $\Delta K_{\mathrm{I}}$-value was derived from each crack line scan (e.g. figure 5.21) from the average slope of the $x$ versus $1/S^2_{\max}$ curves through the two crack-tip regions (see equation (5.8) and related text). As with the crack-length values obtained from these line scans, each $\Delta K_{\mathrm{I}}$-value was associated with the average cycle number, $N$, for the scan. A log–log plot of corresponding $da/dN$ and $\Delta K_{\mathrm{I}}$-values is shown in figure 5.24. The Paris law parameters (equation (5.15)) derived from the intercept and gradient of the best-fit straight line through this data were $C = 0.20 \times 10^{-8}$ mm per cycle and $m = 3.54$. Values given by Smith (1979) for mild steel are $C \simeq 0.25 \times 10^{-8}$ mm per cycle and $m \simeq 3$.

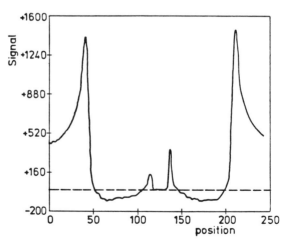

**Figure 5.21** Signal scan along the line of the cracks under mode I loading.

## 5.4 CONCLUSIONS

Crack-tip stress fields under a variety of in-plane, plane-strain loading conditions have been successfully evaluated by means of the thermoelastic stress analysis technique based on the SPATE equipment and valuable comparisons have been made with the results of independent studies. The

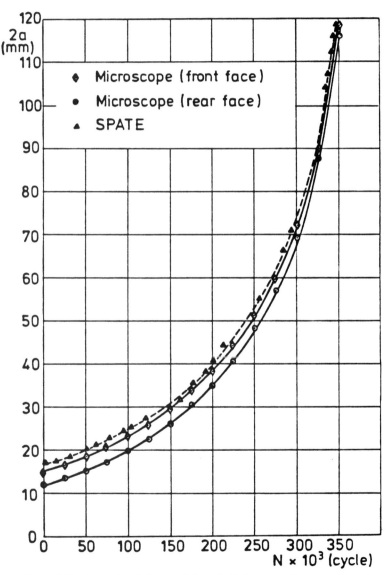

**Figure 5.22** A graph of the crack length (2a) versus the average cycle number, N.

experimental data for single-mode loading work has been processed and presented in such a way that possible errors due to uncertainty in the exact position of the crack (or slot) tip are eliminated and the extent of the region over which the single-term Westergaard equations are valid is readily seen; in particular, the extent of the zone of crack-tip plasticity is clearly indicated.

The work amply illustrates the unusual versatility of the thermoelastic stress analysis technique and the very considerable practical advantages of the technique in relation to specimen preparation, test execution and data processing; major economies are gained in both time and effort. There was no indication in any of the investigations described that the accuracy of the derived stress-intensity factors was in any respect limited by the resolution limits of the equipment. Detailed data at crack-tip level where non-elastic effects may predominate was not forthcoming; work on the thermal effects at the tip may provide a fuller understanding of these effects. There is wide scope for further SPATE applications in the fracture mechanics area; work on

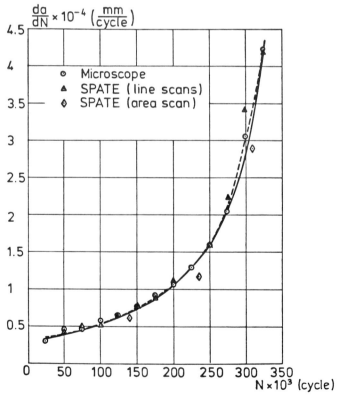

**Figure 5.23** A graph of the crack growth rate versus the average cycle number, *N*.

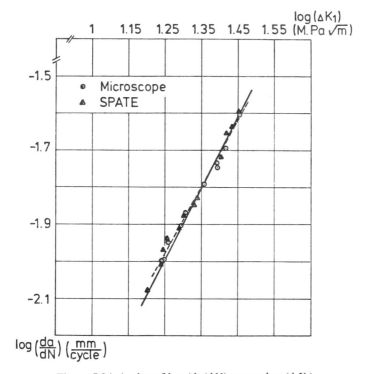

**Figure 5.24** A plot of log (d$a$/d$N$) versus log ($\Delta K_I$).

composites appears to be particularly worthwhile and three-dimensional work of a NDT nature relating to internal cracks in cylindrical vessels holds considerable attraction.

## ACKNOWLEDGMENTS

Figures 5.4 to 5.11 and 5.21 to 5.24 inclusive are reprinted by permission of the Council of the Institution of Mechanical Engineers from the *Proceedings of the International Conference on Fatigue of Engineering Materials and Structures*, 1986, vol 1.

The author is also grateful to:

(i) the Society of Experimental Mechanics for permission to reproduce figures 5.12, 5.17 and 5.18 from the *Proceedings of SEM Spring Conference on Experimental Mechanics, New Orleans, 1986*;

(ii) the International Society of Optical Engineering for permission to reproduce figures 5.13 to 5.16 inclusive from SPIE vol 731, *Proceedings of 2nd International Conference on Stress Analysis by Thermoelastic Techniques, London, 1987*; and

(iii) the Welding Institute for permission to reproduce figures 5.3, 5.19 and 5.20 from *Welding Institute Report* 369/1988.

# 6

# Variable-amplitude Loading
## N Harwood and W M Cummings

## 6.1 INTRODUCTION

As described in earlier chapters of this book, the thermoelastic effect has been exploited successfully for the measurement of full-field surface stress distributions on cyclically loaded structures. The SPATE instrumentation was developed specifically for uniform cyclic loading applications. In practice, most applications of SPATE described in the literature are of a quasi-static nature. Although the SPATE system has proved to be a highly effective tool under quasi-static conditions, using thermoelastic instrumentation only in this way does not fully exploit the potential of a technique which is essentially dynamic in nature.

It was recognised during the development of the SPATE 8000 equipment that the requirement for uniform cyclic loading effectively confines the SPATE system to laboratory environments and thus severely limits its engineering applications. However, the range of application of the technique could be increased greatly if the system could be modified to enable it to handle signals from structures subjected to more complicated loading conditions. Therefore, the National Engineering Laboratory (NEL) undertook to investigate the potential of the SPATE system for random-loading applications as a major part of its thermoelastic stress analysis (TSA) research programme.

Examinations of components under such complex loading conditions have, as expected, shown that the standard instrumentation is incapable of extracting meaningful data using the uniform cyclic signal analysis technique. Structures operating in service are usually subjected to non-periodic, variable-amplitude loading waveforms which the SPATE instrumentation cannot analyse. The main scientific effort of a small team at NEL has been concentrated on research into, and development of, methods to extend the range of TSA to the types of complex, non-periodic signals which are commonly encountered in experimental modal analysis applications and under service-loading conditions.

The use of random-loading signals means that the thermoelastic data contain information covering the entire excitation bandwidth, in contrast to the single-frequency information available from the standard SPATE system. In addition to its implications for service-loading conditions both in the field and in laboratory simulation applications, the wide bandwidth coverage afforded by random excitation is particularly advantageous for modal applications in which the modal density is high, since simultaneous excitation is much more efficient in terms of data acquisition time. Moreover, it is often found that for very lightly damped modes, it is not always possible to obtain complete consistency of excitation over the duration of the scan using a sinusoidal waveform, and in many cases experimental modal analysis is rather limited and time consuming if sinusoidal excitation must be applied at each natural frequency in turn, as in the standard SPATE system.

The fact that an IR radiometer is non-contacting has great advantages for estimating stress in lightly damped structures which are behaving modally, since the act of measurement itself cannot alter the mass distribution of a test structure which may be very sensitive to such changes. Variable shifts in natural frequency produced by a roving, contacting transducer of significant mass make the analysis of measured data much more difficult, particularly using automated procedures. Note that thermoelastic modal data are displayed in the generally more useful form of quantitative, full-field contour maps, rather than the normalised displacement mode shapes produced by traditional experimental modal analysis techniques.

Detailed experimental results published by Cummings and Harwood (1985) on the NEL complex-loading investigations show the feasibility of measuring full-field, random-loading stress patterns using the standard SPATE 8000 system in conjunction with software developed at NEL on a transportable, computer-aided test system. This chapter presents the analytical background on which the NEL random-signal analysis software is based, and experimental thermoelastic data are presented which demonstrate the effectiveness of the technique for the estimation of full-field stress distributions under complex-loading conditions.

## 6.2 SIGNAL ANALYSIS PROCEDURES

This section treats signal processing specifically in the context of the time series analysis techniques used in structural dynamics testing rather than its more general applications.

Originally structural dynamics testing was done using sinusoidal inputs, which have the advantage that all the signal energy is concentrated at a single frequency, giving a high signal-to-noise ratio. It is also a relatively simple technique from an analysis and instrumentation standpoint. However, it has the basic disadvantage that it may take an extremely long time to generate a complete range of mode shapes over a wide bandwidth. This serious drawback led to the development of swept-sine testing, but this technique is also fairly slow due to the necessity of not applying transient pulses and the requirement for sweep rates which do not pass through resonances too rapidly. The inadequacies of sinusoidal techniques and the advent of the digital computer have encouraged the use of random (e.g. white noise) or transient (e.g. impact hammer) signals for most structural dynamics testing applications.

Transient excitation, in which a short impact pulse is applied, the duration of which is related inversely to its bandwidth, is a popular technique for experimental modal analysis applications, and is also a commonly encountered service condition. However, this method would be extremely laborious to use for SPATE applications, due to the requirement for at least one impact—and preferably several—for each pixel position. Nevertheless, the principle of using such a technique for TSA is valid, as may be seen in figure 6.1 which shows a load inpulse and the corresponding thermoelastic response from a steel bar. Transient excitation has also been used at NEL to indicate the irreversible temperature increase associated with the onset of plasticity in an aluminium testpiece (Harwood and Cummings 1989), although the AC coupling of the SPATE detector output complicates the interpretation of the measured data. Future technological developments in fast scanning systems or array detectors will provide the capability of extending TSA to this very important area of applications.

Deterministic signals such as sine waves or impact pulses have an explicit mathematical description. Random signals, however, are non-deterministic, i.e. they cannot be predicted precisely by a mathematical function. Nevertheless, in spite of their unpredictable fluctuations, many random phenomena exhibit some degree of statistical regularity which makes possible a statistical approach to the analysis of such signals. Provided that the signal is ergodic, its statistical invariance allows averaging techniques to be employed to extract useful information from the data, since a single sample record is representative of the complete statistical properties of the signal (Bendat and Piersol 1986). Checks on the invariance of both the RMS of the voltage and the distribution of the frequency content will establish whether a signal is statistically invariant and thus suitable for conventional analysis techniques. Many random signals which are not statistically invariant over extended durations may exhibit sufficient stationarity to allow conventional signal analysis techniques to be applied for limited time periods.

A useful way of representing the amplitude content of a signal is to plot probability or number of counts in a given interval against voltage. Figure 6.2 shows the normal (Gaussian) amplitude distribution typical of many stationary random signals; such signals will also be ergodic. Probability

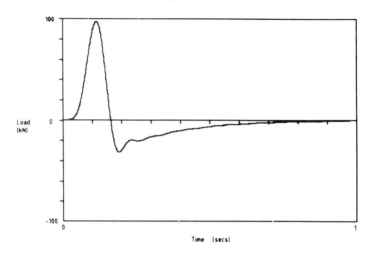

TRANSIENT LOAD PULSE

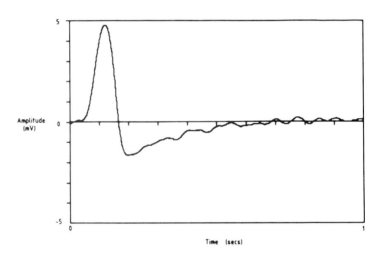

SPATE RESPONSE TO TRANSIENT LOADING

**Figure 6.1** Transient loading data.

distributions can also be used to determine the amplitude characteristics of non-stationary signals.

### 6.2.1 Fourier analysis

Random-signal processing is usually performed in the frequency rather than time domain. Conversion to the frequency domain makes use of the fact that

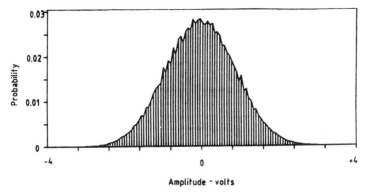

**Figure 6.2** Amplitude probability distribution for random data.

any continuous function $x(t)$ which is periodic in time $T$ can be represented as the summation of an infinite series of sinusoids:

$$x(t) = A_0 + \sum_{n=1}^{\infty} [A_n \cos(2\pi n/T) + B_n \sin(2\pi n/T)]. \tag{6.1}$$

Equation (6.1) is known as a Fourier series for $x(t)$, and $A_n$ and $B_n$ are the Fourier coefficients.

Random signals are, of course, not periodic. However, a quasi-periodicity may be imposed upon a given length of random signal by multiplying it by a window function to weight the signal smoothly towards zero at the edges of the frame.

When acquired by a digital instrument, a continuous analogue signal must be stored as a series of amplitudes at discrete sample points in the time domain. The duration of the sample signal will be given by the number of sample points (i.e. frame size) divided by the sampling frequency ($f_s$).

A function which is defined only at a limited number of discrete points can be represented by a finite Fourier series known as a discrete Fourier transform (DFT). The DFT is usually written in the form

$$S(k) = N^{-1} \sum_{t=0}^{N-1} x(t) \exp(-2\pi i k t / N) \tag{6.2}$$

where $N$ is the number of sample points in each frame of analogue signal, $x(t)$ is the time domain amplitude, and $S(k)$ is the frequency domain complex value ($0 \leqslant k < N/2$). As the above formula shows, $N$ sample points in the time domain yield $N/2$ complex points in the frequency domain.

The above formula may be used directly, particularly to determine the amplitude of a known sinusoidal component amongst random noise, in which case only one spectral line need be evaluated. However, for the fairly large frame sizes normally encountered in signal processing, the calculation is inefficient in terms of computer calculation time if a complete spectrum is to be estimated. Almost invariably in modern signal processing applications a fast-executing algorithm known as the FFT (fast Fourier transform) is used to convert digitised time signals to the frequency domain. The FFT gives a speed improvement ratio of approximately $N/\log_2 N$. The algorithm has the restriction that frame sizes must be an exact power of 2, the most commonly used size in signal analysis being 1024. Thus, the FFT produces a reduction factor in computation time of approximately 100 compared to the DFT for a frame size of 1024. Padding with zeroes or decimation is required to convert data which does not have a power-of-2 segment length into the correct format for the FFT. When a signal has been transformed to the frequency domain, the data will be discretely spaced at frequency intervals (lines) given by the inverse of the duration of the time frame (i.e. $\Delta f = T^{-1}$). It is usually more convenient in FFT-based systems to choose sampling rates which give a spectral resolution which is an exact integer or fraction rather than an unfamiliar decimal number which is not a factor of integer frequencies.

The Fourier transform computes the voltage amplitude (RMS) of the sinusoidal components of a random waveform, expressed as real and imaginary parts. The transform may be produced in a double-sided form containing negative-frequency components. However, since these implicit negative-frequency coefficients are symmetrical with the positive coefficients, they are redundant and are not normally displayed, although their presence must be considered when relating time and frequency-domain amplitudes.

### 6.2.2 Aliasing

It is advisable that the analogue signals are passed through a low-pass filter before being digitised, in order to ensure that there is no significant frequency content in the signal which is higher than the Nyquist frequency (i.e. $f_s/2$). Otherwise aliasing errors will occur produced by frequencies greater than the Nyquist frequency folding back into a lower bandwidth, thus corrupting the data.

A frequency-domain transform contains data up to the Nyquist frequency, but only the bandwidth which is passed unattenuated by the anti-aliasing filter will be usable. It is advantageous to calculate the minimum allowable sampling frequencies relative to the usable bandwidth rather than the Nyquist frequency. The minimum sampling frequencies which ought to be used for common slopes of anti-aliasing filters may be determined by multiplying the maximum frequency of interest in the digitised signal by the factors below:

96 dB/octave — 2.56
72 dB/octave — 3.2

48 dB/octave — 4.

The above data shows the advantage of using as steep an anti-aliasing filter as possible in order to make efficient use of the available computer memory (e.g. the use of a 48 dB/octave filter means that only half of the calculated frequency lines will be usable). In most modern signal analysis systems the filtering is likely to be performed primarily by digital means (i.e. software or firmware) rather than by computer-controlled analogue filters.

The phenomenon of aliasing may be used in a controlled fashion to perform a zoom analysis, i.e. concentrate the available spectral lines in a narrow bandwidth of interest which does not extend down to 0 Hz as would a baseband analysis. The zoom procedure may, however, add a considerable time penalty to the data acquisition, since the formula $\Delta f = T^{-1}$ still applies.

It should be noted that although the sampling frequencies suggested above will determine frequency content correctly, they do not necessarily give an accurate representation of the peak amplitudes of an analogue signal. For an accurate discretised time-domain representation, a sampling frequency of at least eight times the maximum frequency of interest in the signal is required.

### 6.2.3 Leakage and windowing

As mentioned previously, the Fourier transform requires that the data be exactly periodic in the time frame and therefore a weighting window is applied to remove the abrupt truncation of the signal at the edges of the frame and impose a quasi-periodicity on the data. Signals which are Fourier transformed when not weighted in this way will include in the transform the frequency components necessary to account for the apparent discontinuities at the edges of the frame. These spurious data are known as leakage errors since energy effectively 'leaks' from actual frequencies to adjacent lines, thus smearing the data. Note that the weighting requirement also applies to periodic signals which happen not to be exactly periodic in a particular time frame.

Many modern signal analysis systems are fitted with an integral digital-to-analogue facility which allows the operator to output repeated bursts of random signal with a duration which is identical to that of the data acquisition time frame. This facility may be used to produce pseudo- or periodic-random excitation, which has the advantage that there ought to be no leakage errors, since the application of a weighting function is not required to ensure periodicity.

The Hanning cosine bell is the most popular window for continuous random signals. It is equivalent to multiplying the time-domain frame by the function $w_f$ below:

$$w_f = [1 - \cos(2\pi t/T)]/2. \tag{6.3}$$

The use of this window does not eliminate the effect of leakage entirely, but the roll-off occurs at 18 dB/octave rather than the 6 dB/octave for unwindowed data. The energy attenuation produced by the imposition of the Hanning window may be compensated for by multiplying the function by the factor 1.633. Many other forms of window are available for signal processing: the most suitable window for the estimation of the amplitudes of fairly clean sinusoids is the 'flat top', which produces a negligible amplitude error, even when the sinusoid does not fall exactly on one of the computed spectral lines; for transient applications a window which suppresses amplitudes at the start of a frame is unsuitable, and therefore in order to ensure that there is no leakage error due to truncation, mathematical damping is usually added to the response by applying an exponential taper if there is a significant vibration remaining at the end of the frame. For modal applications this extra damping must be subtracted before modal damping is estimated.

### 6.2.4  Overlap processing

In order to obtain a result which may be used with confidence, the averaging of several frames must be performed. The speed of modern digital computers is such that the FFT processing time may be considerably less than is necessary to acquire the time histories from structures operating at the lower frequency range commonly encountered in service conditions. In such a case, the use of double buffering (i.e. simultaneous data acquisition and computation) means that more efficient use of the computer is achieved by segmenting time frames so that the next frame to be processed overlaps a section of the previous frame rather than being sequential with it; the amount of overlap may be adjusted so that the processing time approaches the acquisition time for the new data in each frame of time signal.

Although data processed in this way cannot have the same statistical properties as completely independent samples, the process is more effective than using all the data points just once, particularly with windowed data which contain points that are severely attenuated in one frame but that will be restored to virtually full amplitude in a subsequent overlapped frame. Off-line overlap processing may also be performed from a series of stored sequential time histories.

Most of the attenuated data produced by the application of a Hanning window can be recovered using an overlap factor of two (i.e. 50% of the frame size). There is a slight further advantage in using an overlap factor of four, since this produces a uniform overall weighting effect (Randall 1987). Overlap factors of eight or more offer few significant further advantages.

### 6.2.5  Power spectra

The Fourier transform process produces an array of complex (i.e. real and imaginary) coefficients for each time frame, which gives detailed information relevant to the Fourier analysis process. However, mathematical data in this form are not usually of much interest to engineers, and therefore a self-conjugate multiplication is often performed in order to convert the data to a simpler form known as an auto-power spectrum. An auto-power spectrum, which is a real rather than a complex function, describes the frequency composition of a signal in terms of its mean-square amplitudes. In modal applications it is particularly important to check the auto-power spectrum of the force input so that the spectral distribution of the forcing energy is known.

The joint properties of two simultaneous signals may also be expressed in terms of a cross-power spectrum, in which case the graph displays the power common to both signals and the phase difference between them over the required bandwidth. Auto- and cross-power spectra are computed as shown below:

$$G_{xx} = n^{-1} \sum_1^n S_x S_x^*$$

$$G_{yy} = n^{-1} \sum_1^n S_y S_y^* \qquad (6.4)$$

$$G_{yx} = n^{-1} \sum_1^n S_y S_x^*$$

where $S_x$ is the Fourier transform of the reference signal, $S_y$ is the Fourier transform of the response signal, $G_{xx}$ is the auto-power spectrum of the reference signal, $G_{yy}$ is the auto-power spectrum of the response signal, $G_{yx}$ is the cross-power spectrum between the reference and the response, and $n$ is the number of ensembled averages. * denotes the complex conjugate. The error on an auto-power spectrum estimate at a spectral line is proportional to $n^{-1/2}$.

The cross-power spectrum is rarely computed as an end in itself, but rather as an intermediate step in the calculation of frequency response and coherence functions. As may be seen in the above expression for the error band on an auto-power spectrum estimate, frequency-domain averaging must be performed in the computation of power spectra in order to give progressively more dependable, smoother results. The number of averages used is a compromise between the generation of accurate estimates and the length of time required to collect the data. Note that the expression for the error band only refers to random, not systematic, errors. Continuous power spectra are often expressed as density functions by a simple normalisation relative to the spacing of the spectral lines, thus producing amplitudes which are independent of the frame size used for the data acquisition.

A sensitive method of establishing whether a signal contains periodic components is to calculate an auto-correlation function. This function should only show significant correlation at zero time for wide-band random data, whereas signals containing periodic components will have an auto-correlation function which persists over all time values. An auto-correlation function may be computed by performing an inverse Fourier transform on an auto-power spectrum, although a 'circularity' error will be produced unless a zero-insertion technique is used to compensate for this (Otnes and Enochson 1978).

### 6.2.6 Frequency response functions

Under modal conditions there is an interaction between mass, stiffness, and damping forces which may produce considerable structural amplification of the forcing function at natural frequencies of the structure. A convenient form of displaying resonant behaviour is to compute an ensemble-averaged ratio between the response and the forcing function at a given point on the structure as a function of frequency. This frequency-domain gain and phase plot, which is known as a transfer or frequency response function, gives vector information about the dynamic behaviour of the structure over the excitation bandwidth. On many dynamic systems, there is a direct linear relationship between input and output, characterised by a set of frequency response functions (FRFS).

As an example of a frequency response function measured on an actual structure, figure 6.3 shows the gain and phase relationship between the acceleration response from a point on the floor of a vehicle and the reference

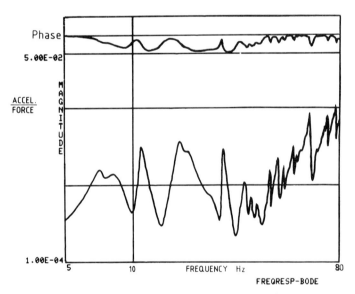

**Figure 6.3** Sample frequency response function.

force. The plot takes the form of a compound graph displaying both the gain and the phase between the reference and response signals over a given bandwidth. Resonances are indicated by local peaks in the frequency response function gain together with an associated phase change. Resonant frequencies which lie within the bandwidth of service vibrations may present potential difficulties to a designer or engineer.

The measurement of frequency response functions is a popular means of indicating the dynamic behaviour of structures since the noise reduction is considerably better than can be achieved by single-channel functions, and the peaks in the function reflect modal behaviour of the actual structure rather than irregularities in the forcing function; i.e. it isolates the intrinsic dynamic characteristics of the structure from any particular input signal.

Several different ways of calculating frequency response functions have been developed over recent years in order to improve their accuracy in the presence of noise (Rocklin *et al* 1984). The results of all the methods coincide in a noise-free environment, but when noise is present, the various estimators will give different values.

The traditional way of computing a frequency response function for the standard single input–output configuration is to divide the cross-power spectrum by the auto-power spectrum of the forcing function. This is usually written in the form

$$H_1 = G_{yx}/G_{xx}. \tag{6.5}$$

However, it is often found that this traditional frequency response function estimator (i.e. $H_1$) does not give an accurate estimate around very lightly damped resonances, since the force usually falls to very low levels at the natural frequency, and thus the signal-to-noise ratio may be very poor. In such a case, provided that the noise level in the response is fairly low, it would probably be advantageous to compute the inverse frequency response function ($H_2$) by dividing the auto-power spectrum of the response by the complex conjugate of the cross-power spectrum. This procedure may be expressed mathematically in the form

$$H_2 = G_{yy}/G_{xy}. \tag{6.6}$$

$H_2$ tends to be less sensitive than $H_1$ to leakage errors. Provided that leakage errors are not significant and that noise on the reference and response signals is uncorrelated, $H_1$ and $H_2$ must form the lower and upper bounds, respectively, for the range within which the actual frequency response function must lie.

Considerable efforts have recently been made to find a frequency response function estimator with optimal characteristics (Mitchell *et al* 1987). An estimator known as $H_v$ has been developed for multi-point excitation conditions. The computation of this estimator involves noise-vector matrix manipulations which require considerable user effort and relatively long computational times. Under single-input conditions, however, the $H_v$ estimator reduces to the geometric mean of $H_1$ and $H_2$, and it may therefore be used as a general estimator which will give an acceptable accuracy over a wide range of modal conditions and which, unlike $H_1$ and $H_2$, has uniform properties at resonances and anti-resonances. $H_v$ is particularly suited to conditions in which there is significant noise on both the input and output signals around natural frequencies. $H_v$ is computed by dividing the cross-power spectrum by its own absolute value and multiplying the result by the square root of the ratio of the response to the reference auto-power spectra, as shown below:

$$H_v = \frac{G_{yx}}{|G_{yx}|}\sqrt{\frac{G_{yy}}{G_{xx}}} = \sqrt{H_1 H_2}. \qquad (6.7)$$

Note that $H_1$, $H_2$ and $H_v$ all have identical phase values preserved in the cross-spectrum term; only the gains may vary.

The three estimators mentioned previously are each calculated from a combination of the cross-power and auto-power spectra. Since a cross spectrum is effectively a correlated function, it is insensitive to incoherent noise. However, measured auto-power spectra tend to have a greater amplitude than the actual value due to the addition of noise, which is why $H_1$ and $H_2$ are low- and high-biased estimators, respectively.

The fact that auto-spectra are prone to noise errors, even when averaged considerably, has led to the development of an 'unbiased' estimator, known as $H^c$, which is computed entirely from cross-spectra. This estimator does, however, have the disadvantage that it requires an extra data acquisition channel for every excitation position. The extra signal which must be connected is the function generator or power amplifier output which is the source of the excitation. $H^c$ is computed by dividing the cross-power spectrum between the response and the drive signal by that between the force and the drive signal, as shown below:

$$H^c = G_{ys}/G_{xs}. \qquad (6.8)$$

It has been claimed (Mitchell *et al* 1987) that for many modal conditions the $H^c$ estimator tends to be more accurate than $H_1$ or $H_2$ when the coherence is low. $H^c$ is liable to vary in both gain and phase from the other three estimators discussed earlier.

Frequency response functions should strictly only be calculated for linear systems with constant parameters, i.e. the basic statistical properties of the system must be invariant with time and the functions must be independent of the amplitude (within limits) of the excitation signal. However, the ensemble averaging of data generated by an infinitely variable, random excitation signal tends to mask any non-linearities which may be present in the structure.

### 6.2.7 Coherence

It is useful to be able to assess the quality of any particular frequency response function measurement which is being made from ensemble-averaged, pure random data. The function normally used for such assessment is known as the coherence function, and it gives an indication of the dependence between two signals, usually the measured response and the load input reference signal. The coherence function is effectively a frequency-domain correlation function; its value should always lie between 0 (no correlation) and 1 (complete correlation or perfect causality); spurious values greater than 1 are sometimes generated in digital systems when division by numbers close to zero is performed. The value of the coherence at any given frequency will indicate how many ensembled averages will be needed to give a reliable estimate of the frequency response function value at that frequency (Randall 1987). It should be noted that the coherence function itself will appear noisy and tend towards too high a value if insufficient ensemble averaging was performed in its calculation; at least 20 averages should be taken to give a reliable estimate.

The coherence function ($\gamma^2$) is computed by dividing the square of the magnitude of the cross-power spectrum by the product of the input and output auto-power spectra, i.e.

$$\gamma^2 = \frac{|G_{yx}|^2}{G_{xx}G_{yy}}. \qquad (6.9)$$

The symbol for coherence is usually given as a squared term to make clear that it is the counterpart of the square of the correlation coefficient used in statistics and to avoid confusion with the occasional description of the coherence function as being the positive square root of the above value. Provided that noise in the measured output can be assumed to be the only significant factor influencing coherence, multiplication of the output power spectrum by the coherence function will show the portion of the output power which is due to the input function, and the signal-to-noise ratio in the output may be determined by dividing $\gamma^2$ by $(1-\gamma^2)$.

The traditional and inverse frequency response functions are also related via the coherence function, as shown below:

$$\gamma^2 = H_1/H_2. \qquad (6.10)$$

High coherence implies reliability in the measured frequency response function estimates, even when the individual power spectra appear noisy. However, it must be remembered that spurious high coherence can sometimes be produced by cross-talk in the instrumentation when very low-amplitude signals, such as an IR detector output, are being acquired. Moreover, high coherence can be produced when signals which are not completely random are being analysed. The coherence function will only be truly indicative of the quality of the measurement if the signals are truly random from average to average.

The coherence function is often used simply as a general indicator of noise contamination. However, a more sophisticated interpretation of the possible causes of low coherence may be made by reference to the list below:

(i) extraneous noise is present in one of the signals or uncorrelated noise in both signals;

(ii) the system is non-linear, i.e. the damping or stiffness is changing significantly with the amplitude of the excitation;

(iii) the output is being partially excited by another input signal;

(iv) there is excessive leakage due to the wrong choice of window function;

(v) there is a time lag between the signals of the same order as the duration of a frame; or

(vi) a bias error has been produced by the use of too coarse a frequency resolution at frequencies around which the data are changing rapidly.

In case (i) above, a low-coherence value does not necessarily mean that an incorrect FRF estimate is being made at that frequency, but it does indicate that the measurement is a difficult one and thus a large number of ensemble averages ought to be performed to give confidence in the result. Note that the systematic nature of the low-coherence situations in the remaining cases listed above means that the quality of the FRF data is insensitive to averaging.

It is good experimental practice to make regular measurements of coherence in order to monitor the quality of the calculated FRFs. This allows any deterioration in the measurement conditions, such as would be produced by instrumentation malfunction or coupling from an incoherent source, to be detected quickly.

The coherence function corresponding to the FRF shown in figure 6.3 is displayed in figure 6.4. For modal analysis applications it is desirable to have high coherence around resonant frequencies; low-coherence values elsewhere are less important since in most applications these data need not be used to estimate the modal contribution.

## 6.3 MODAL ANALYSIS

Modal analysis is a procedure for investigating structural vibration, in which the dynamic behaviour of an elastic structure, which is being excited by a forcing function, is described in terms of mode shapes at resonant frequencies. Most noise or vibration problems are related to resonance phenomena, where service conditions excite one or more modes of vibration. In addition to attempts to reduce vibration as much as possible at source (e.g. by proper

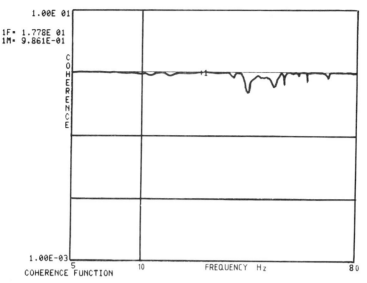

**Figure 6.4** Sample coherence function of figure 3.

balancing), rotating machinery may be isolated on spring mountings of appropriate stiffness. Extra damping may also be added in the form of coatings or lumped damping elements, or structures may be placed on a base made from a resilient, energy-absorbing material. However, a basic principle of structural design is that resonant and excitation frequencies should, if possible, not be close to one another.

Associated with each resonant (natural) frequency is a particular spatial displacement or deformation shape, usually called a mode shape or eigenvector. Most structures which can be considered to be constant-parameter linear systems within the amplitude of the excitation may be adequately described just in terms of their modes of vibration. In addition to the mode shape at each natural frequency, there are associated values of the modal mass, modal stiffness, and modal damping. Thus, many complicated structures can be considered to have an overall dynamic representation equivalent to a series of simple, single-degree-of-freedom oscillators, each consisting of a mass, a spring, and a damper element. The modes therefore form a basis for a complete dynamic description of a structure.

Modal analysis of components may be performed using purely theoretical techniques such as the FE method, but for complicated structures it would be advisable to use data from experimental measurements to refine the analytical model in order to confirm that the mathematical representation of the structure together with its loading and boundary conditions are realistic. If an accurate mathematical model can be developed, its use may shorten the lead time between initial design and final prototype, since design changes may be made and assessed in the computer without the need to go through so many expensive and laborious construction phases required for a series of development prototypes.

The evaluation of the dynamic behaviour of structures makes a vital contribution to the design process, since static analysis techniques are inappropriate for modal conditions. Modern requirements for improved product performance have encouraged the increasing use of modal analysis of industrial components and structures which are subjected to dynamic service conditions.

Experimental modal analysis in its modern, sophisticated form is a computer-aided-testing (CAT) technique which aids the design of structurally efficient, cost-effective and reliable products for dynamic service conditions. Structural dynamics testing of prototypes and finished products is vital for locating assembly defects, assessing structural integrity, and identifying vibration problems or damaging modal behaviour. The use of a modal survey in a trouble-shooting role provides vital feedback which may be used to facilitate design improvements and to predict how best to control or optimise the dynamic behaviour of a structure in its operating environment.

A commonly performed experimental modal analysis technique requires that the structure be shaken at a single point by a random drive signal (usually white noise) which covers the excitation bandwidth to which the structure is subjected in service. A load cell attached between the forcing device (i.e. a servo-hydraulic actuator or electromagnetic shaker) and the structure provides a known reference against which the dynamic response of the structure is measured. An idealised geometry representing the relative Cartesian coordinates of the response points is generated in the computer. The response is usually measured by a roving triaxial accelerometer which is placed at each geometry point in turn. An alternative simple technique, commonly used in field applications, involves the use of an impact hammer to provide the excitation.

Specialist software may be used to analyse the structural frequency response functions generated between the reference and response signals. Several distinct methods of curve-fitting the experimental data are available depending on the nature of the damping and the coupling of the modes. Arrays of modal coefficients describing the dynamic behaviour at each resonance are gradually built-up as the analysis procedure progresses. When these arrays are complete, engineering assessments may be made by display-ing the mode shapes dynamically at various angles on a visual display unit to simulate the vibration in slow motion.

When a mode shape has been determined experimentally, the dynamic behaviour of the structure at that resonance may be simplified to that of an equivalent single-degree-of-freedom model by generating a mode enhance-ment function. This function represents a linear combination of all the FRFs in the set, each weighted according to the size of the appropriate mode shape vector. The compliance in a viscously lightly damped, single-degree-of-freedom system will reach an amplitude at resonance of approximately $\frac{1}{2}(K\zeta)^{-1}$ where $K$ is the modal stiffness and $\zeta$ is the viscous damping ratio. The compliance may be determined from inertance (i.e. acceleration-based) data by double integration. The bandwidth around the natural frequency in the mode enhancement function may be curve-fitted to estimate damping. Once the damping ratio and the compliance amplitude at resonance have been measured, an estimate of the modal stiffness may then be calculated. When the modal stiffness is known, the modal mass may be determined by dividing the stiffness by $(2\pi f_r)^2$, where $f_r$ is the natural frequency in hertz. The modal mass and stiffness are often calculated experimentally in order to supply parameters to, or compare data with, theoretical models. However, it must be borne in mind that the calculated value for the modal mass is entirely dependent on how the mode shape was scaled. Therefore, in order to enable a unique value for the modal mass to be determined, a mode shape should be normalised in a standard fashion. This is often done by setting the largest modal coefficient or the vector at the reference location to unity before the mode enhancement function is generated.

An experimental modal analysis may be validated by using the database to synthesize predicted FRFs for a different excitation position. These analytically predicted functions may then be compared with new experimental data. The effects of simple structural modifications or changes in excitation can be predicted from the modal database. The confirmed modal parameters may subsequently be used to develop or refine a FE model of the structure, the use of which can be extended to simulate alternative configurations of loading or more sophisticated physical modifications. Provided that the experimental and theoretical mode shapes can be produced with an identical mesh geometry, they may be compared mathematically and discrepancies located.

## 6.4 APPLICATION TO THERMOELASTIC STRESS ANALYSIS

The frequency response function is used as the basis of the thermoelastic stress measurement technique for random applications. These functions are computed using the thermoelastic response signal instead of the displacement or acceleration response which is generally used in experimental modal analysis. The relationships between the time series functions described in this report are shown in figure 6.5.

The basic validity for the analysis of thermoelastic response signals acquired from randomly loaded structures is demonstrated in figure 6.6 which shows a load cell output and the simultaneous SPATE signal from a flat steel bar subjected to a 1–40 Hz white-noise drive signal in a servo-hydraulic test machine. The correlation between the signals can be seen very clearly

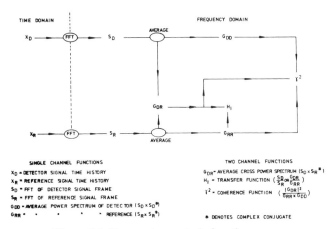

**Figure 6.5** Time series analysis functions.

when the thermoelastic response has been low-pass filtered at 40 Hz to remove the high-frequency noise and mains harmonics which are invariably present in the SPATE output signal. Note that in figure 6.6 the phase of the response signal has been reversed to show the correlation clearly.

A suite of software has been written at NEL on a GenRad 2515 computer-aided test system in a specialized language for vibration analysis called TSL (Time Series Language). TSL uses control instructions which are similar to BASIC, but the language has been specifically designed for signal processing and the rapid analysis of large blocks of data. Source programmes may be precompiled before execution so that they run much more rapidly than when under the control of a statement-by-statement, interpretive compiler. There is an extensive library of callable machine-code subroutines for all the arithmetic computations related to Fourier transforms and spectrum averaging, which can easily be incorporated in the high-level language software, and which will execute at very high speed. The GenRad 2515 is equipped with a 4-channel data acquisition facility and programmable digital anti-aliasing filters that can be used up to a maximum signal frequency of 25 kHz. The system also contains a digital-to-analogue output channel which may be used to synchronise the data acquisition with the movement of the scanning mirrors within the SPATE head, as indicated schematically in figure 6.7.

For simple specimens loaded under white-noise conditions, the absolute value $(a^2 + b^2)^{1/2}$ of the complex FRF value $a + ib$ at a given frequency has been found to be proportional to the sum of the principal stresses at that frequency, and the phase $(\tan^{-1} b/a)$ indicates the polarity of the stress relative to the reference signal. For quasi-static applications, the FRF data may be bandwidth averaged over the pre-resonant frequency range in order to obtain maximum noise reduction.

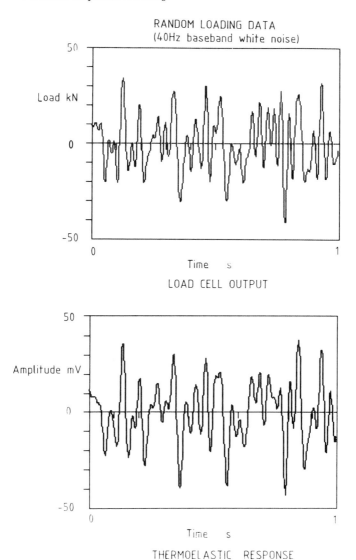

**Figure 6.6** Random loading time-domain data.

Three examples of the practical use of the NEL software are described below. Each component was loaded in a servo-hydraulic test machine driven by a 4–40 Hz white-noise drive signal. The chosen dwell-time was 1 s/point in each case, i.e. a similar value to that commonly used in sinusoidal applications.

Figure 6.8 shows a relief map representing the distribution of the sum of the principal stresses measured around a hole in an axially loaded plate. The

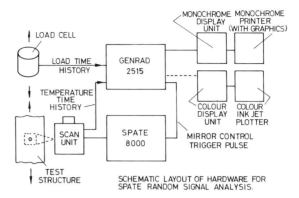

**Figure 6.7** Schematic of random-signal analysis hardware.

measured stress distribution agrees closely with the theoretical solution (Timoshenko and Goodier 1982).

Figure 6.9 displays the stress pattern measured on a flat chain-link 480 mm in length. This investigation formed part of a study to improve the fatigue performance of large chain-links for offshore applications. The stress concentrations around the internal corners of the link can be seen clearly in the random-loading pattern.

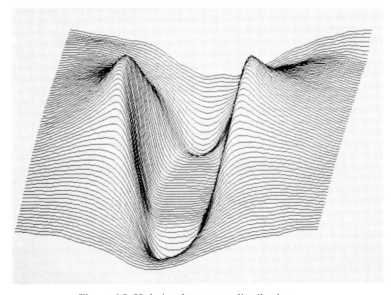

**Figure 6.8** Hole-in-plate stress distribution.

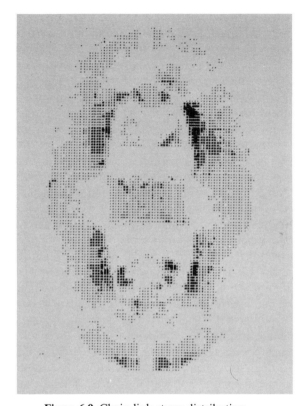

**Figure 6.9** Chain-link stress distribution.

Both of the above components were loaded using a load cell output as a reference signal. However, under service conditions a load cell attached to the structure is unlikely to be available. In such a case, a strain gauge would normally provide the most appropriate method of obtaining a reliable reference signal. Stress patterns have been acquired from a welded cruciform joint using a strain gauge as a reference. This work has been described by Harwood and Cummings (1989). A two-dimensional version of the cruciform joint was also manufactured and tested under random loading (Cummings and Harwood 1987). These components were tested as part of a wider programme investigating the fatigue behaviour of offshore structures.

The above stress patterns are displayed in monochrome, but the stored raw data can easily be transferred and copied in conventional colour-coded format when the transportable hardware is connected to a colour-display workstation. The quality of the stress patterns acquired at NEL under random loading is similar to that which would be expected for sinusoidal conditions, despite the more difficult noise-rejection problems encountered in random-signal analysis applications.

A digital analysis technique similar to that used for random signals has also been developed for sinusoidal-loading conditions as part of an investigation into possible means of reducing the capital cost of thermoelastic stress analysis equipment. Stress pattern data produced by this digital technique appeared to show much less edge-effect than that produced by the standard SPATE analogue instrumentation (Harwood and Cummings 1989).

Finite element modelling using the ANSYS software package has been carried out on several of the structures examined using SPATE under sinusoidal- or random-loading. The two main aims of this study were to separate principal stress vectors using a verified model for the sum of the principal stresses and to encourage the direct inclusion of SPATE experimental data in the design procedures for structures and components. These aspects are described fully in Chapter 9.

For modal applications the standard procedure is to describe the dynamic behaviour of a system in terms of a series of natural frequencies and mode shapes. Conventional experimental modal analysis techniques generate mode shapes in terms of arrays of displacement coefficients on an idealised structure with a much coarser description than that normally used for a FE mesh. Although it is theoretically possible to calculate stress/strain distributions from experimentally generated displacement data, such data would normally contain insufficient degrees-of-freedom to enable stresses to be estimated reliably, even if a computerized differentiation facility was available on the analysis system. SPATE, therefore, has great potential for experimental modal analysis applications, since a database may be generated in terms of quantitative stress values, rather than the generally less useful normalized displacement displays produced by traditional techniques. The use of sensitive IR equipment to measure the thermoelastic response enables full-field data to be generated as a matrix of closely spaced stress scalars at each natural frequency. It should be noted that the experimental modal velocity data produced by the VPI system mentioned in Chapter 1 may have fine enough spatial resolution for the estimation of rotations and for comparison with FE meshes, although single-axis data may be somewhat inconvenient for the analysis of three-dimensional structures. The use of the VPI system for experimental modal analysis has been described by Bream *et al* (1989).

A basic assumption of all modal analysis methods is that, in the vicinity of resonance, the predominant response is due to the contribution of the mode with the closest natural frequency. The simplest method of measuring modal stresses is to take the gain and phase at the FRF peak as the modal contribution. This method is known as the peak-amplitude or total-response technique, and it assumes that all the response at the natural frequency can be attributed to a single mode, whereas other analysis methods take into account contributions which may be present from other modes. The FFT-based frequency response function, being a discrete rather than a continuous function, means that the use of a frequency-line spacing which is too coarse will smear sharp resonances and thus produce an inaccurate estimate of the gain at lightly damped natural frequencies. For FRFs which exhibit well-separated modes, the peak-amplitude method could be expected to produce acceptable accuracy. However, in many cases, particularly where high damping is present, this simple technique would be unsuitable since it gives no isolation from the residues of closely coupled modes, and thus should not be used in cases where there is a significant amount of modal interaction.

Another simple method of estimating the modal contribution when there is a slight amount of coupling is to neglect the real part of the complex FRF magnitude and to use only the imaginary component. This method, which is known as quadrature response, is based on the fact that the imaginary component approaches zero away from the resonant frequency. This technique has been built-in to the suite of NEL software together with the overlap processing and zoom facilities described earlier.

A more sophisticated technique for the separating-out of the contributions of fairly closely coupled modes involves the curve-fitting of circles to a narrow bandwidth around a resonance. It is based on the fact that a plot of the real component against the imaginary component of a FRF will trace a circular arc for each resonance. The technique is known as Nyquist plotting. Such a Nyquist plot for thermoelastic data, measured on a stiffened plate over a narrow bandwidth around a flapping mode, is shown in figure 6.10; a least-squares circular curve-fit algorithm has been used to draw the circle through the data. Figure 6.10 shows that this particular mode is completely uncoupled. For lightly coupled modes the circle is displaced slightly from the origin, and for more heavily coupled modes only an arc of a circle for each mode will be generated. The frequency limits of each arc must then be determined by the operator before the circle-fitting subroutine is applied to

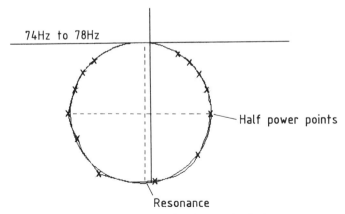

**Figure 6.10** Nyquist plot for thermoelastic data.

each arc in turn in order to separate the modal contributions. It is advisable to have at least two data points above the half-power points for a reliable circle fit to be obtained.

For a normal mode, the modal contribution is a scalar, the amplitude of which is given by the diameter of the circle, and its polarity is determined by whether the centre of the circle is above or below the resonant frequency point. The natural frequency is estimated from an interpolation to determine the point on the circle at which the rate of change of frequency is a maximum. If the circle is tilted significantly to the imaginary axis, this indicates that non-proportional damping has caused the mode to become complex, i.e. all the points in the mode shape are not always exactly in- or out-of-phase with each other. The Nyquist technique is very effective at isolating the effects of nearby modes, provided that the coupling is not so great that the separate arcs may not be distinguished, but unfortunately the technique does have a strong tendency to diverge from sensible values when the circles are ill-defined, possibly leading to poor results under purely automatic operation.

Once the natural frequency $(f_r)$ and the half-power points $(f_1, f_2)$ have been identified, damping may be estimated from the formula

$$\zeta = \frac{f_1 - f_2}{If_r}. \qquad (6.11)$$

If the damping mechanism is viscous, $I = 2$; for light, hysteretic damping, $I = 1$.

In practice, many engineering structures exhibit damping characteristics that may be a combination of a wide range of effects such as viscous, structural and, possibly, radiation or frictional (coulomb) damping. The above formula expresses damping non-dimensionally in terms of critical damping (Thomson 1972), i.e. $\zeta = \frac{1}{2}C(KM)^{-1/2}$ as shown in the following equation of motion for a viscously damped, single-degree-of-freedom system:

$$M\ddot{u} + C\dot{u} + Ku = F(t) \qquad (6.12)$$

where $M$ is the mass, $\ddot{u}$ is the acceleration, $C$ is the damping constant, $\dot{u}$ is the velocity, $K$ is the stiffness, $u$ is the displacement, and $F(t)$ is the excitation force.

Measurements at NEL indicate that damping estimates made from curve-fitting a narrow frequency-band around structural resonances on thermo-elastic FRFs produce very similar values to that generated from functions determined using an accelerometer.

For cases where the degree of coupling is so great that modal contributions cannot be separated using the so-called single-degree-of-freedom techniques described above, more complicated multi-degree-of-freedom equations, which define the complete mathematical behaviour of one or more FRFs over a bandwidth, may be used to curve-fit the data (Ewins 1984). At NEL the thermoelastic data may be interfaced with a powerful, commercial modal analysis package (MPLUS) which contains the latest developments in multi-degree-of-freedom modal parameter estimation techniques and which resides on the same computer system.

## 6.5 CALIBRATION

Calibration under random-loading conditions is determined in terms of the RMS amplitude, since this is statistically constant for a stationary signal. The calibration procedure is essentially similar to that used for the sinusoidal excitation as described in Chapter 1, except that values are computed in the frequency domain rather than the time domain. The standard technique is to measure the time-domain RMS amplitude of the sum of the principal stresses using a strain-gauge rosette bonded to a chosen calibration position on the structure. As may be seen from a Mohr's circle, the power spectrum of the summed outputs of the two gauges which are normal to each other will give an output which is invariant with gauge direction. Thus the gain and phase information in the FRFs will be measured relative to a scalar rather than a vector quantity. The shape of the combined spectrum of the orthogonal gauges shows how the energy content of the signals is distributed over the excitation bandwidth. The thermoelastic response around the gauge area must then be related to the amplitude of this strain spectrum over the excitation bandwidth. This is done by measuring a thermoelastic FRF at the calibration position. The gain of this function is then divided by the measured strain spectrum to produce a calibration function which may subsequently be applied to convert other thermoelastic FRFs to quantitative strain-values for a scan performed on a linear system. Strain may be converted to stress using known values of Young's modulus and Poisson's ratio. The measured phase values are used to determine the polarities of the thermoelastic stresses.

Note that this FRF calibration technique relies on the excitation signal remaining ergodic throughout the calibration procedure and during the scan time. This ergodicity requirement is a particularly important consideration for SPATE applications, since the instrument scans fairly slowly, point-by-point, rather than producing an instantaneous image. The use of a four-channel ADC, such as that incorporated in the computer-aided test system used at NEL for thermoelastic random-excitation applications, means that two extra channels are available for the acquisition of strain gauge data which may be used to check for stationarity or otherwise aid in the calibration of the FRF data.

The thermoelastic data are, of course, in the form of the sum of the principal stresses. The principal stresses may, however, be separated experimentally using a frictional strain-gauge probe pressed onto the surface in three directions at each point of interest, as described in Chapter 1.

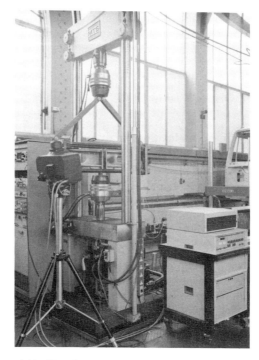

**Figure 6.11** Chassis cross-member during SPATE analysis.

## 6.6 TEST RESULTS FOR MODAL CONDITIONS

An experimental modal analysis was performed on a stiffened plate using a conventional technique, i.e. a fixed force gauge and a roving accelerometer. A thermoelastic stress analysis was then performed on the plate under random-loading conditions. The excitation bandwidth was restricted to 50–150 Hz in order to improve the signal-to-noise ratio. Problems with low coherence at natural frequencies meant that not all of the resonant stress patterns excited within the above bandwidth could be extracted reliably from the thermo-elastic data generated by the wide-band random drive signal. However, the NEL software successfully generated two distinct stress patterns from thermoelastic FRFs stored during the course of a single scan (Harwood and Cummings 1987). For cases of wide-band random-excitation of lightly damped structures it would be an advantage to shape the spectrum of the drive signal so that energy could be concentrated at the frequencies of interest, and thus improve the sigal-to-noise ratio where the most important measurements are to be made.

Further tests have been performed on a vehicle chassis cross-member. Pre-resonant stress patterns were measured on the end-cap under sinusoidal conditions (Loader *et al* 1987). The structure in its loading frame is shown during SPATE analysis in figure 6.11. Attachment of an accelerometer to an end-cap allowed the natural frequencies of the structure to be located (e.g. figure 6.12). Random loading was then applied, and frequency response functions were measured from a highly stressed point on the end-cap. A sample modal stress pattern acquired under wide-band random excitation is shown in figure 6.13. Six modes were excited simultaneously by a 40 Hz baseband white-noise signal. However, the most highly stressed part of the end-cap shown in the stress pattern was relatively unaffected by the modal behaviour of the overall structure below 40 Hz. Therefore, all the stress patterns extracted by the NEL software had a similar form (Harwood and Cummings 1989).

A theoretical modal analysis was performed on a cantilever using the ANSYS software. The dimensions of the cantilever were adjusted until three structural modes were predicted to lie below 40 Hz. The cantilever was then manufactured and an experimental modal analysis performed using an electromagnetic shaker to determine experimentally the natural frequencies and shapes for the first three modes. These mode shapes are displayed in figure 6.14. SPATE was then used to determine stress patterns at each of the natural frequencies in turn. The scan area was restricted to the base of the cantilever around two bolt-heads since the stresses remote from these fixtures were found to be very low. A frequency response function was then determined between the thermoelastic response adjacent to a bolt-head and the input force reference. This thermoelastic FRF (figure 6.15) showed the resonant frequencies clearly. A SPATE scan was then performed under 40 Hz baseband white-noise excitation using the NEL software as described previously. Stress patterns were extracted from the stored FRFs at each of the

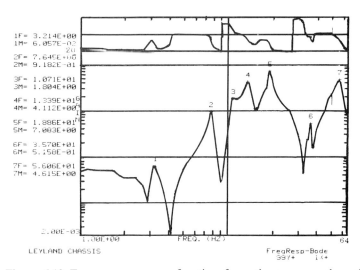

**Figure 6.12** Frequency response function from the cross-member of figure 6.11.

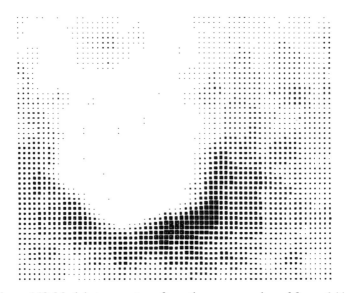

**Figure 6.13** Modal stress pattern from the cross-member of figure 6.11.

three natural frequencies. The simultaneously excited stress patterns produced by random loading (figure 6.16) compared well with the data measured
by the standard SPATE system using sinusoidal excitation at each frequency in
turn (Harwood and Cummings 1989). Random-loading data do tend to be
noisier than the sinusoidal, but the advantages of random loading are clearly
apparent in situations where there is a high modal-density within the

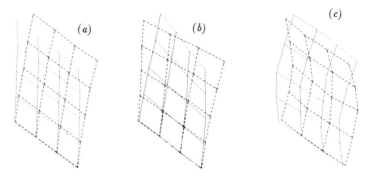

**Figure 6.14** Mode shapes from a cantilever at the three natural frequencies: (*a*) 6 Hz; (*b*) 16 Hz; and (*c*) 34 Hz.

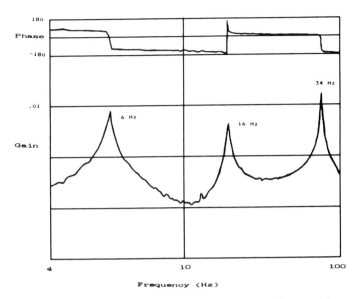

**Figure 6.15** Thermoelastic FRF of the cantilever of figure 6.14.

bandwidth of interest and for cases in which consistent excitation of lightly damped modes is difficult to achieve using a sinusoidal waveform.

Thermoelastic modal data acquired at NEL confirmed that $H_1$ tended to produce a low estimate at lightly damped resonances due to the poor signal-to-noise ratio in the reference, whereas $H_2$ usually produced a high estimate due to the high noise level which is invariably present in the response; $H_2$ was also sensitive to the presence of mains harmonics in the SPATE output. The $H_v$ estimator generally appeared to be significantly smoother than $H_1$ or $H_2$ for the same number of averages, although it was more sensitive than $H_1$ to the presence of mains harmonics in the SPATE signal. $H^c$ may be the most appropriate estimator if an extra channel is available and if very few ensemble averages are being taken.

It is very important to use an estimator which can be readily smoothed in few averages, since a point-by-point measurement technique such as SPATE could take an excessively long time to generate a stress pattern if a long dwell-time per point has been used in order to give more confidence in the accuracy of the stress patterns. Results indicate that $H^c$ removes bias and is superior to $H_v$ at rejecting the mains harmonics in the SPATE signal, but that averaging is not effective at smoothing the $H^c$ estimator. Note that the effectiveness of $H^c$ in rejecting mains harmonics will only apply if the load signal is relatively free of mains harmonics, since correlated signal components are not rejected by a cross-spectrum. Test results obtained at NEL indicate that, in general, $H_v$ appears to be the most suitable estimator for modal stress analysis using the

SPATE system on lightly damped structures, particularly if estimates are not being made at frequencies close to mains harmonics.


## 6.7 DISCUSSION AND CONCLUSIONS


This chapter has described signal processing techniques which are widely used for the analysis of complex waveforms. These techniques have been applied to infra-red thermoelastic response signals from dynamically loaded structures, with the aim of measuring full-field stress distributions produced by random excitation.

The random-loading investigations at NEL have indicated that the thermoelastic technique has great potential as a unique method of measuring full-field stress distributions on structures operating under non-periodic loading conditions, including modal behaviour. A hardware/software system

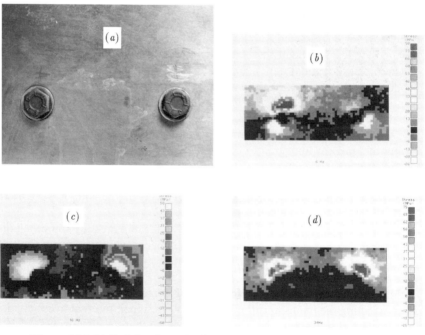

**Figure 6.16** (*a*) Bolt-heads at the base of the cantilever of figure 6.14; (*b*) random-loading stress pattern extracted at the 6 Hz spectral line; (*c*) random-loading stress pattern extracted at the 16 Hz spectral line; and (*d*) random-loading stress pattern extracted at the 34 Hz spectral line.

has been developed which can be combined with the standard SPATE 8000 to form a transportable system which is capable of analysing random-loading signals from structures operating in service.

The system has been found to produce excellent data under quasi-static random loading where bandwidth averaging can be applied to improve the noise reduction. The greatest potential of the technique, however, lies in its application to modal conditions, where stored data from a single scan can subsequently be analysed to extract a range of stress patterns at all resonant or anti-resonant frequencies which lie within the excitation bandwidth. The ability to generate a modal database has great potential for the experimental verification of dynamic finite-element models. Frequency response function measurements between the thermoelastic response from a stiffened plate and a wide-band random forcing function indicate that commonly used single-degree-of-freedom modal analysis techniques may be employed to generate a modal stress database from thermoelastic response and random-excitation force data. The wide bandwidth coverage produced by random excitation is in contrast to the single-frequency information available from the standard SPATE equipment. The SPATE system may be used to determine normal modal stresses, although its use would be very laborious in cases where the modal density is high. In the case of complex modes, the SPATE 8000 is unsuitable since a single phase-lock angle between the reference and response signals is assumed to be valid over the entire scan area.

In addition to some of the more obvious advantages of the use of wide-band random excitation for experimental stress analysis which have been described previously in this chapter, the regular measurement of frequency response functions is a sensitive method of detecting any significant structural changes in a component which is behaving modally. Such phenomena as crack growth, delamination or loosening fixtures will change the stiffness of the structure and will thus progressively shift resonant frequencies. Modal testing at NEL has shown that thermoelastic stress analysis under random loading can be used to detect such changes in dynamic behaviour. Thus the capability of the stress measurement system discussed in this chapter to identify resonant frequencies has proved to be a very useful extension of currently available full-field experimental stress analysis and NDT techniques.

Random forcing functions may have the disadvantage that adequate excitation of all the modes can sometimes be difficult to achieve using white-noise excitation on a lightly damped structure. This problem could be minimised by using an equaliser to shape the spectrum of the drive signal to concentrate its energy around the frequencies of interest. Nevertheless, the estimation of modal stresses under random loading is bound to be a more difficult measurement situation than the quasi-static condition, since bandwidth averaging cannot be performed to reduce noise, and at frequencies above 30 Hz the thermal properties of commonly used paint coatings modify the measured response. Experimental investigations at NEL indicate that $H_v$

is the most suitable FRF estimator for modal conditions in which the signal-to-noise ratio of the reference signal is poor at the frequencies of interest.

A disadvantage of the FRF technique for the estimation of quantitative stress values is that the calibration procedure must be repeated if the excitation amplitude or the source of the reference signal is changed. The accuracy of the FRF technique in producing quantitative stress patterns depends on the statistical properties of the reference signal being invariant. Any change in the excitation level will cause an equivalent change in the overall thermoelastic response. However, since a FRF is the ratio of a response to a reference, the changes will cancel and the function will remain the same, despite the stress having changed. Therefore, the calibration procedure, which relates a FRF to a 'known' stress, would have to be repeated. In the uniform cyclic signal technique this situation does not arise, since the amplitude of the reference signal is irrelevant (within limits) to the result of the analysis procedure. Investigations at NEL into the use of the response signal alone for experimental stress analysis under random loading revealed that the data were too noisy to enable measurements to be made of acceptable repeatability. A single-channel technique could also not be used if phase polarity information was required, and such a technique would be much less suitable for measurements of modal behaviour, particularly when the form of the forcing function is not known.

Although the thermoelastic data acquired under random loading is likely to be somewhat noisier than can be achieved under sinusoidal conditions, the pattern-smoothing software which has been developed for the random-signal analysis system allows the stress data to be presented in a form which may be virtually indistinguishable in quality to sinusoidal patterns measured on the standard SPATE equipment.

Calibration software has been written to determine quantitative 'sum of the principal stress values' from the random data, although the calibration procedure is inevitably somewhat more complicated than that required under sinusoidal conditions. The use of a frictional strain-gauge probe and associated software may allow principal stress vectors to be separated at key points on the surface of the structure.

## ACKNOWLEDGMENT

This chapter is published by permission of the Chief Executive, National Engineering Laboratory, Department of Trade and Industry. It is Crown copyright.

# 7

# The Thermoelastic Effect—A Higher-order Consideration

## A K Wong

## 7.1 INTRODUCTION

Although the coupling between mechanical deformation and the thermal energy of an elastic body was noted as early as 1805 by Gough, it was Lord Kelvin who brought about the understanding of this physical phenomenon by his theoretical treatment in 1853. To this day, it is virtually impossible to talk about the thermoelastic effect without referring to what is known as Kelvin's Law, namely

$$\delta T/T = -K_0 \delta\sigma \qquad (7.1)$$

in which $T$ is the absolute temperature of the element under consideration, $\delta T$ is the change in temperature, $\delta\sigma$ is the change in the sum of the principal stresses, and $K_0$ is the thermoelastic constant.

For small temperature changes, it is generally assumed that $K_0$ is a material constant which is independent of the applied stress. As SPATE was designed to respond linearly with changes in temperature, this means that given a fixed stress amplitude, the SPATE signal should be independent of any applied static load. However, the experimental work of Machin et al (1987) revealed that this was in fact not the case. It was shown that for the titanium and aluminium alloy specimens used, the SPATE signal appeared clearly to be linearly dependent on the mean-stress applied. The phenomenon of a stress-dependent thermoelastic response had previously been observed by Belgen (1968), although it had completely escaped the attention of later researchers.

Belgen conjectured that the apparent dependence of $K_0$ on stress was attributable to the effects of stress on the specific heat and/or Poisson's ratio. But without any real physical basis to account for this phenomenon, the findings of Machin *et al* (1987) were generally not taken seriously amongst SPATE users. In the same year, Wong *et al* (1987) presented a revised thermoelastic theory which showed that the inclusion of higher-order terms could indeed explain the observed anomaly. This, together with later works by Wong *et al* (1988a, b), quickly dispelled any scepticism and initiated growing interest as the possibility of exploiting this phenomenon for measuring residual stresses was demonstrated.

This chapter consolidates the theoretical and experimental investigations carried out in the Aeronautical Research Laboratory, Australia, and presents a physical interpretation of the mean-stress effect. The extension of the theory to the analysis of anisotropic materials is also presented and the need for a non-adiabatic approach briefly described.

## 7.2 THEORY

For the sake of clarity, the following derivation of the thermoelastic equation will be restricted to isotropic materials without any internal heat source. The extension to anisotropic materials is trivial and will be presented in a later section. The conservative laws which govern the mechanics of *small* quasi-static deformations are:

(i) Conservation of mass

$$\rho = \rho_0 \tag{7.2}$$

where $\rho$ and $\rho_0$ are the strained and unstrained densities of the material, respectively.

(ii) Conservation of momentum

$$\sigma_{ij,j} = -\rho_0 F_i \tag{7.3}$$

summing over $j$: $j = 1$–$3$, and where $\sigma_{ij}$ is the stress tensor and $F_i$ is the body force per unit mass.

(iii) Conservation of energy

$$\rho_0 \dot{U} = \sigma_{ij} \dot{\varepsilon}_{ij} - Q_{i,i} \tag{7.4}$$

summing over $i, j$: $i, j = 1$–$3$, and where $\dot{U}$ is the rate of change of internal energy per unit mass, $\dot{\varepsilon}_{ij}$ is the strain rate, and $Q_i$ is the heat flux through a unit surface area whose outward directed normal is in the direction of $x_i$.

Finally, the material response is required. For an isotropic Hookean material, this takes the form:

(iv) Constitutive law

$$\sigma_{ij} = 2\mu\varepsilon_{ij} + (\lambda\varepsilon_{\kappa\kappa} - \beta\delta T)\delta_{ij} \tag{7.5}$$

in which

$$\beta = (3\lambda + 2\mu)\alpha \qquad \delta T = T - T_0 \tag{7.6}$$

where $\alpha$ is the coefficient of linear thermal expansion, $\lambda$ and $\mu$ are the Lamé constants, $T_0$ is the reference temperature, $\delta_{ij}$ is the Kronecker delta, and $\varepsilon_{\kappa\kappa}$ is the first strain invariant. Equation (7.5) is often referred to as the Duhamel–Neumann Law.

Introducing the Helmholtz free energy function, $\Phi$, namely

$$\Phi = U - TS \tag{7.7}$$

in which $S$ is the specific entropy, and selecting $\varepsilon_{ij}$ and $T$ to be the independent state variables, we have

$$\dot{\Phi} = \frac{\partial\Phi}{\partial\varepsilon_{ij}}\dot{\varepsilon}_{ij} + \frac{\partial\Phi}{\partial T}\dot{T} = \dot{U} - T\dot{S} - S\dot{T} \tag{7.8}$$

or

$$\dot{U} = \frac{\partial\Phi}{\partial\varepsilon_{ij}}\dot{\varepsilon}_{ij} + \frac{\partial\Phi}{\partial T}\dot{T} + T\dot{S} + \dot{T}S. \tag{7.9}$$

Substituting equation (7.9) into equation (7.4) yields

$$\left(\rho_0\frac{\partial\Phi}{\partial\varepsilon_{ij}} - \sigma_{ij}\right)\dot{\varepsilon}_{ij} + \rho_0\left(\frac{\partial\Phi}{\partial T} + S\right)\dot{T} + \left(\rho_0 T\dot{S} + Q_{i,i}\right) = 0. \tag{7.10}$$

For a reversible process, the second law of thermodynamics requires that

$$Q_{i,i} = -\rho_0 T\dot{S}. \tag{7.11}$$

As this leads to the vanishing of the third term in equation (7.10), and also that the terms within the remaining parentheses are independent of $\dot{\varepsilon}_{ij}$ and $\dot{T}$, it may be concluded that

$$\sigma_{ij} = \rho_0\frac{\partial\Phi}{\partial\varepsilon_{ij}} \tag{7.12}$$

and

$$S = -\frac{\partial\Phi}{\partial T}. \tag{7.13}$$

From equations (7.12) and (7.13), it follows that

$$\dot{S} = -\frac{\partial^2\Phi}{\partial\varepsilon_{ij}\partial T}\dot{\varepsilon}_{ij} - \frac{\partial^2\Phi}{\partial T^2}\dot{T}$$

$$= -\frac{1}{\rho_0}\frac{\partial\sigma_{ij}}{\partial T}\dot{\varepsilon}_{ij} - \frac{\partial^2\Phi}{\partial T^2}\dot{T}. \tag{7.14}$$

The substitution of equation (7.14) into equation (7.11) gives

$$Q_{i,i} = T\left(\frac{\partial\sigma_{ij}}{\partial T}\dot{\varepsilon}_{ij} + \rho_0\frac{\partial^2\Phi}{\partial T^2}\dot{T}\right). \tag{7.15}$$

It is customary to define a specific heat under constant strain $C_\varepsilon$ such that for $\dot{\varepsilon}_{ij} = 0$

$$\rho_0 C_\varepsilon \dot{T} = -Q_{i,i} \tag{7.16}$$

which by setting $\dot{\varepsilon}_{ij} = 0$ in equation (7.15) we get

$$C_\varepsilon = -\frac{\partial^2\Phi}{\partial T^2} \tag{7.17}$$

or that equation (7.15) may be written as

$$Q_{i,i} = T\frac{\partial\sigma_{ij}}{\partial T}\dot{\varepsilon}_{ij} - \rho_0 C_\varepsilon \dot{T}. \tag{7.18}$$

From the constitutive law, equation (7.5), we obtain

$$\frac{\partial\sigma_{ij}}{\partial T} = 2\frac{\partial\mu}{\partial T}\varepsilon_{ij} + \left(\frac{\partial\lambda}{\partial T}\varepsilon_{\kappa\kappa} - \frac{\partial\beta}{\partial T}\delta T - \beta\right)\delta_{ij}. \tag{7.19}$$

Note that in the present derivation the material properties are assumed to be general functions of temperature. This differs from Kelvin's approach in which the material properties were assumed to be constants. The present approach is reasonable as it is well known that such properties can be strong functions of temperature.

The substitution of equation (7.19) into equation (7.18) yields

$$Q_{i,i} = T\left[\left(-\beta - \frac{\partial\beta}{\partial T}\delta T + \frac{\partial\lambda}{\partial T}\varepsilon_{\kappa\kappa}\right)\delta_{ij} + 2\frac{\partial\mu}{\partial T}\varepsilon_{ij}\right]\dot{\varepsilon}_{ij} - \rho_0 C_\varepsilon \dot{T}. \tag{7.20}$$

In the case of a stress-induced temperature field in which $\delta T$ is extremely small, an order of magnitude analysis shows that whilst $(\partial\beta/\partial T)\delta T$ may be considered negligible compared to $\beta$, the terms $(\partial\mu/\partial T)\varepsilon_{ij}$ and $(\partial\lambda/\partial T)\varepsilon_{\kappa\kappa}$ can be of significant order. Hence, omitting the former term and assuming adiabatic conditions (i.e. $Q_{i,i} = 0$), we have

$$\rho_0 C_\varepsilon\frac{\dot{T}}{T} = -\left(\beta - \frac{\partial\lambda}{\partial T}\varepsilon_{\kappa\kappa}\right)\dot{\varepsilon}_{\kappa\kappa} + 2\frac{\partial\mu}{\partial T}\varepsilon_{ij}\dot{\varepsilon}_{ij} \tag{7.21}$$

or in terms of stresses,

$$\rho_0 C_\varepsilon\frac{\dot{T}}{T} = -\left[\alpha + \left(\frac{v}{E^2}\frac{\partial E}{\partial T} - \frac{1}{E}\frac{\partial v}{\partial T}\right)\sigma_{\kappa\kappa}\right]\dot{\sigma}_{\kappa\kappa} + \left(\frac{(1+v)}{E^2}\frac{\partial E}{\partial T} - \frac{1}{E}\frac{\partial v}{\partial T}\right)\sigma_{ij}\dot{\sigma}_{ij} \tag{7.22}$$

where $E$ is Young's modulus, $v$ is Poisson's ratio, and $\sigma_{\kappa\kappa}$ is the first stress invariant.

From equation (7.22) it can be seen that the revised formulation shows the rate of temperature change to be a function of both the stresses and their rate of change. To see how this may affect the SPATE response, consider a one-dimensional stress state of the form

$$\sigma_{11} = \sigma_{\kappa\kappa} = \sigma_m + \delta\sigma \sin \omega t$$

$$\sigma_{22} = \sigma_{33} = \sigma_{12} = \sigma_{13} = \sigma_{23} = 0. \tag{7.23}$$

The substitution of equation (7.23) into equation (7.22) yields

$$\rho_0 C_\varepsilon \frac{\dot{T}}{T} = -\left(\alpha - \frac{1}{E^2}\frac{\partial E}{\partial T}(\sigma_m + \delta\sigma \sin \omega t)\right)\omega \,\delta\sigma \cos \omega t$$

$$= -\left(\alpha - \frac{1}{E^2}\frac{\partial E}{\partial T}\sigma_m\right)\omega \,\delta\sigma \cos \omega t + \frac{1}{2E^2}\frac{\partial E}{\partial T}\omega(\delta\sigma)^2 \sin 2\omega t. \tag{7.24}$$

Integrating equation (7.24), assuming $\delta T = T - T_0 \ll T_0$ and noting that energy transfer to the surroundings would ensure a vanishing static component in $\delta T$, we have

$$\rho_0 C_\varepsilon \frac{\delta T}{T_0} = -\left(\alpha - \frac{1}{E^2}\frac{\partial E}{\partial T}\sigma_m\right)\delta\sigma \sin \omega t - \frac{1}{4E^2}\frac{\partial E}{\partial T}(\delta\sigma)^2 \cos 2\omega t. \tag{7.25}$$

The difference between this response and that which is predicted by Kelvin's Law may be seen by neglecting the temperature-dependent terms in equation (7.25), namely

$$\rho_0 C_\varepsilon \frac{\delta T}{T_0} = -\alpha \,\delta\sigma \sin \omega t. \tag{7.26}$$

## 7.3 QUANTITATIVE VALIDATION

Equation (7.25) represents the temperature response to an applied sinusoidal load. One immediate observation is that it is composed of two harmonic components. The first is at the applied loading frequency, $\omega$, and the corresponding amplitude is a function of both $\sigma_m$ and $\delta\sigma$. On the other hand, the second harmonic component (of frequency $2\omega$) is dependent only on the square of the stress amplitude $(\delta\sigma)^2$. Although the coefficient $(1/4E^2)(\partial E/\partial T)$ is expected to be comparatively small, this component can become significant when the applied stress amplitudes are sufficiently large. However, as far as SPATE is concerned, only the fundamental component is detected as signals from all other frequencies are effectively filtered out by the correlator. Consequently, the amplitude of the temperature response as seen by SPATE is

$$\frac{\delta T}{T_0} = -\frac{1}{\rho_0 C_\varepsilon}\left(\alpha - \frac{1}{E^2}\frac{\partial E}{\partial T}\sigma_m\right)\delta\sigma. \tag{7.27}$$

Comparing this to Kelvin's Law, equation (7.1), we see that the revised formulation yields an effective thermoelastic parameter which is mean-stress dependent, namely

$$K_e = \frac{1}{\rho_0 C_\varepsilon}\left(\alpha - \frac{1}{E^2}\frac{\partial E}{\partial T}\sigma_m\right). \tag{7.28}$$

A normalised measure of the dependence of $K_e$ on $\sigma_m$ is thus

$$\frac{1}{K_0}\frac{\partial K_e}{\partial \sigma_m} = -\frac{1}{\alpha E^2}\frac{\partial E}{\partial T} \tag{7.29}$$

where $K_0 = \alpha/(\rho_0 C_\varepsilon)$ is the conventional thermoelastic constant.

Interestingly, equation (7.29) shows that the stress dependence of the thermoelastic parameter arises as a result of the temperature dependence of the material's stiffness as opposed to Belgen's conjecture on the stress dependence of the specific heat or Poisson's ratio. A physical interpretation of this effect is presented later in section 7.5.

To see just how well the present theory can explain the observed mean-stress dependence of $K_e$, equation (7.29) may be applied once the properties $\alpha$, $E$ and $\partial E/\partial T$ are known. Although $\partial E/\partial T$ data are not as well studied as for other material properties, such data are available for the titanium alloy Ti-6Al-4V (see *Metals Handbook* 1980) and the aluminium alloy Al-2024 (see Brammer and Percival 1970) used in the earlier experiments. This provided Wong *et al* (1987) with a direct test on their revised theory. Table 7.1 lists the data used and the comparison between the theoretically predicted and measured mean-stress dependence for these materials. Excellent agree-

**Table 7.1** Comparison of the theoretical and measured mean-stress dependence of $K_e$.

| Material | $\alpha$ (°C$^{-1}$) | $E$ (MPa) | $\partial E/\partial T$ (MPa °C$^{-1}$) | $(\partial K_e/\partial \sigma_m)K_0^{-1}$ (MPa$^{-1}$) Theory, equation (7.29) | Experiment[a] |
|---|---|---|---|---|---|
| Ti-6Al-4V | $9.0 \times 10^{-6}$ | $1.11 \times 10^5$ | $-48.0$ | $4.33 \times 10^4$ | $4.29 \times 10^{-4}$ |
| Al-2024 | $2.3 \times 10^{-5}$ | $7.2 \times 10^4$ | $-36.0$ | $3.02 \times 10^{-4}$ | $3.19 \times 10^{-4}$ |

[a] Machin *et al* (1987).

ment between theory and experiment is clearly evident, thereby strongly supporting the validity of the theory presented.

Whilst we are classifying such a phenomenon as a higher-order effect, it is interesting to note that, under suitable conditions, such an effect could become significant. Take the titanium case for instance; it may be seen from the above result that $K_e$ can vary by as much as 43% over a mean-stress range of 1000 MPa. Such a mean-stress range spanning tension and compression can easily be accommodated as the yield strength of this material is around 900 MPa. Other materials which have a high temperature-dependent stiffness might also be susceptible to this mean-stress effect and this must therefore be accounted for when interpreting SPATE results.

## 7.4 FURTHER EVIDENCE

The foregoing analysis showed that by including the temperature-dependent terms, the thermoelastic parameter becomes mean-stress dependent. Another interesting aspect which became apparent is that the thermoelastic response now consists of two harmonic components. From equation (7.25) it can be seen that for a loading frequency of $\omega$, the temperature response, besides having a primary component at the loading frequency, should also contain a component at $2\omega$. Experiments demonstrating this effect were carried out by Wong et al (1988a). Bypassing SPATE's standard correlator, an FFT spectrum analyser was used to monitor the raw IR detector signal. The same unit was also used to monitor the loading waveform, making sure that an essentially pure sine wave was being applied. A series of tests covering a stress amplitude range of approximately 100–400 MPa and a frequency range of 5–20 Hz were performed on a Ti-6Al-4V specimen. Indeed, it was found that with a single applied harmonic load, a small but detectable signal at the second harmonic frequency in the SPATE detector signal was observed for the range of tests considered.

Figures 7.1 and 7.2 show the measured amplitude of the first and the second harmonic responses, respectively. It can be seen that the first harmonic signal is linearly related to the applied stress amplitude and the slope of the line of best fit represents the conventional thermoelastic constant. On the other hand, the second harmonic amplitude was, as predicted by the revised theory, found to be proportional to the square of the stress amplitude. To remove the need to calibrate the measurements, the ratio of the slopes of the $2\omega$ and $\omega$ results was taken, giving a value of $1.13 \times 10^{-4}$ MPa$^{-1}$. This compares exceptionally well with the theoretically predicted result, where it may be shown that

$$\frac{\text{slope}(2\omega)}{\text{slope}(\omega)} = \left| \frac{1}{4\alpha E^2} \frac{\partial E}{\partial T} \right| = 1.08 \times 10^{-4} \text{ MPa}^{-1} \tag{7.30}$$

for the aluminium alloy used. Such good agreement further serves to validate the revised thermoelastic theory.

Experiments were also undertaken in Wong *et al* (1988a) to show the general quadratic nature of the thermoelastic response. By using a correlation technique, the general response law was extracted from experimental data and the mean-stress dependence of $K_e$ was derived. This has the advantage that the loading waveform does not need to be purely sinusoidal. The results for the aluminium and titanium alloys used, as well as that for the 4340 steel presented in Dunn *et al* (1989), again agreed well with the theoretical predictions.

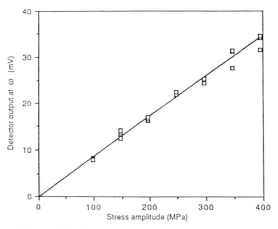

**Figure 7.1** First harmonic detector output.

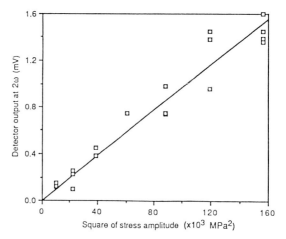

**Figure 7.2** Second harmonic detector output.

## 7.5 A PHYSICAL INTERPRETATION

Whilst the revised theory clearly explains the observed phenomenon, the question on the physical meaning of this effect is invariably raised. This is best answered by considering the following example as shown in figure 7.3. A bar of uniform cross-sectional area is fixed vertically at the upper end and a mass is attached to its free lower end. Assuming the mass gives rise to a uniform stress $\sigma$ throughout the bar, the expression relating the strain $\varepsilon$ to $\sigma$ and an applied temperature change $\delta T = T - T_0$ is given by Hooke's Law, namely

$$\varepsilon = (\sigma/E) + \alpha(T - T_0) \tag{7.31}$$

where $T_0$ is a reference temperature at which the thermal strain is considered to be zero.

Since the definition for the coefficient of linear thermal expansion, $\alpha$, is

$$\alpha = \frac{\partial \varepsilon}{\partial T}\bigg|_{\sigma = \text{constant}} \tag{7.32}$$

then equation (7.31) is strictly correct only for the case where the Young's modulus $E$ is independent of temperature. In reality, $E$ for most materials is a function of temperature. Furthermore, the measurement of the coefficient of thermal expansion in practice is invariably taken at zero-stress. Hence, a more appropriate description would be

$$\varepsilon = (\sigma/E) + \alpha_0(T - T_0) \tag{7.33}$$

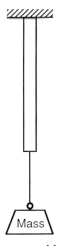

**Figure 7.3** Vertically mounted bar with hanging mass.

where $\alpha_0$ is given by

$$\alpha_0 = \left. \frac{\partial \varepsilon}{\partial T} \right|_{\sigma = 0} \tag{7.34}$$

and represents the measured values which are reported in material data references. If we now apply equation (7.32), and assume $E$ to be a general function of temperature, we have

$$\alpha = \alpha_0 - \frac{1}{E^2} \frac{\partial E}{\partial T} \sigma. \tag{7.35}$$

Equation (7.35) reveals that the actual coefficient of thermal expansion is stress dependent. In this case, the thermal expansion process can be viewed as being composed of two parts. Assuming both $\alpha_0$ and the temperature change $\delta T$ to be positive for instance, the first part, associated with $\alpha_0$, represents the extension of the bar due to the increase in the mean interatomic-spacing as a result of increased thermal excitation. Secondly, the bar will further change in length as a result of the change in its stiffness when a stress is present. From figure 7.3 it can be easily visualised that if the bar softens due to a temperature rise, the hanging mass will cause the bar to extend downwards by an amount which is proportional to the weight attached. The latter phenomenon is attributed to the stress-dependent nature of the coefficient of thermal expansion and thus explains the stress dependence of the thermo-elastic parameter. This is easily seen by taking $K_e = \alpha/\rho_0 C_\varepsilon$, so that

$$\frac{\partial K_e}{\partial \sigma} = \frac{1}{\rho_0 C_\varepsilon} \frac{\partial \alpha}{\partial \sigma} = -\frac{1}{\rho_0 C_\varepsilon E^2} \frac{\partial E}{\partial T} \tag{7.36}$$

which is the identical result as may be obtained by the more general derivation presented in section 7.3.

## 7.6 RESIDUAL STRESS MEASUREMENTS

The significance of establishing an understanding of the mean-stress effect is two-fold. One immediate consequence is that this effect must be taken into account when using SPATE or any other thermoelastic technique, particularly when a high static-stress is applied and/or when the material's elastic properties are strongly temperature-sensitive. Secondly, and perhaps more importantly, the potential for using the thermoelastic effect to detect residual stresses in a component is now apparent, as residual stresses are basically static (or mean) stresses. In a classic experiment, Wong *et al* (1988b) demonstrated this possibility by successfully using SPATE to measure the residual stress distribution of several aluminium bars which had undergone prior plastic deformation.

Starting with specimens which were purposely machined-curved, the specimens were straightened in a four-point loading rig, thereby introducing the classical N-shaped residual stress profiles in the middle region (see figure 7.4). A series of strain gauges bonded across the specimen prior to the straightening process allowed these profiles to be determined and were used to compare with the SPATE results. A dynamic load was then applied in the longitudinal direction of the specimens and SPATE scans were taken over the middle region. For the purpose of comparison, a SPATE scan was also taken on a geometrically similar specimen which had not been deformed plastically and was therefore assumed to contain no residual stresses. Figures 7.5 and 7.6 (Plate 3) show the comparison of the scans of the 'straight' and 'straightened' specimens. As the dynamic stresses were identical for both of these specimens, namely a uniformly distributed uniaxial dynamic stress, the application of Kelvin's Law would have predicted identical scans. However, it can be clearly seen that the specimen which had undergone the plastic

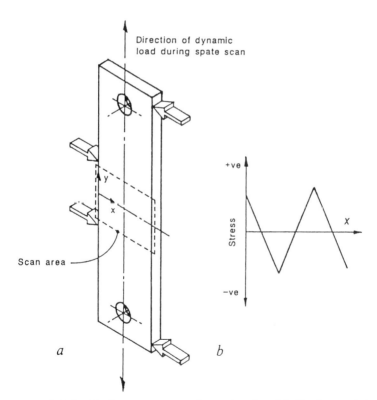

**Figure 7.4** Residual stress test specimen showing (*a*) the four-point loading configuration, and (*b*) a typical residual stress profile across the width of the specimen.

deformation showed a significantly different result from the uniformly scattered pattern of the unworked specimen.

Due to the mean-stress effect, in which the presence of a tensile static stress produces a larger SPATE signal whilst a compressive static stress has the opposite effect, the four distinct bands in figure 7.6 in effect highlight the two tensile and two compressive residual stress regions which are typical of such specimens. As the scan data appeared to vary essentially only across the width of the specimen, the data were collapsed vertically to form a line scan. Figure 7.7 shows the comparison of the residual stress profile derived from the SPATE data with that of the strain-gauge data. The overall agreement between the two sets of results may be considered good, although some discrepancies were found at the region where the material had undergone compressive plastic deformation (left-hand side of the figure). It was pointed out in Wong *et al* (1988b) that similar discrepancies were reported by workers in acoustoelastic residual stress measurements (see Fisher and Herrmann 1984, Hirao and Pao 1985) where the consistently underpredicted results in this region are thought to be due to the change in physical properties of the material as a result of compressive yielding. But in any case, this result represents the first demonstration of the application of the thermoelastic effect for measuring residual stresses within a material.

In the above example, the residual stress field was essentially one-dimensional and the procedure used for determining the residual stresses was relatively simple. Tackling a general two-dimensional problem would be much more difficult and, as yet, no successful attempt has been achieved. Recently, however, Ryall *et al* (1990) have shown that the non-linearity in the thermoelastic response can indeed be exploited to yield individual stress

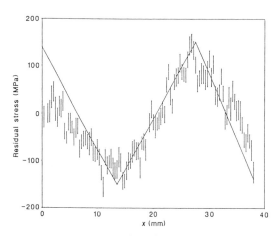

**Figure 7.7**  Residual stress profile for a four-point straightened specimen. The vertical bars indicate the standard error for the averaged SPATE data and the continuous line represents the predicted results based on the strain-gauge data.

components of both the dynamic and static stresses. Using actual SPATE data for a plate containing a circular hole which was loaded with both a mean and a dynamic force, and by imposing the conditions of equilibrium and compatibility, they were able to determine each of the stress-component fields. Current effort is being devoted to the determination of the residual stress field around a cold-expanded bolt hole.

## 7.7 IMPLICATIONS FOR COMPOSITE MATERIALS

For completeness, a discussion of the thermoelastic behaviour of composites, particularly on those aspects which are usually neglected, is briefly presented here. For a more detailed discussion, the reader is referred to Chapter 8.

Although the higher-order effect has been clearly demonstrated for metals, its consideration for composite materials is equally, if not more, important. Because the stiffness of such materials can be highly temperature-sensitive, an intuitive extension of the previous analysis to composites tells us this would give rise to a strongly stress-dependent thermoelastic behaviour. This is indeed the case. For an orthotropic material, the general equation governing the thermoelastic effect is given by Jones $et\ al$ (1987), namely

$$-Q_{i,i} + \rho_0 q \dot{M} = \rho_0 C_\varepsilon \dot{T} - \{C_{ijkl,T} \varepsilon_{kl} - [\beta_{ij}(T - T_0)$$
$$+ \Upsilon_{ij}(M - M_0)]_{,T}\} T_0 \dot{\varepsilon}_{ij} \qquad (7.37)$$

where $C_{ijkl}$ is the elasticity tensor, $Q_i$ is the heat flux rate, $q$ is the heat generated per unit mass due to moisture absorption, $M$ is the moisture content per unit mass, $\Upsilon_{ij} = \psi_{kl} C_{ijkl}$ in which $\psi_{kl}$ are the coefficients of moisture expansion, and $\beta_{ij} = \alpha_{kl} C_{ijkl}$ in which $\alpha_{kl}$ are the coefficients of thermal expansion. Under normal testing conditions, the moisture absorption terms may be neglected as the diffusion of moisture is a relatively slow process. However, as mentioned earlier, terms involving the dependence of the elasticity tensor on $T$ cannot, in general, be neglected. As in the case for metals, it can be seen from equation (7.37) that this leads to the sensitivity of the temperature response to an applied mean-stress (or mean-strain as this is a more convenient variable for the case of composites). Moreover, since the stiffness of composites is much more dependent on temperature than it is for metals, the thermoelastic responses of composites are expected to be more mean-strain sensitive. In the experiment of Dunn $et\ al$ (1989), it was shown that the thermoelastic parameter for the $\pm 45$ degree graphite/epoxy specimen used exhibited a mean-strain dependence of approximately 17% per 1000 microstrain. This is indeed a significant effect considering such a specimen has a strain capacity in excess of 5000 microstrain!

Whilst the mean-strain effect in composites is important, its prediction from theory is made extremely difficult by virtue of the many more material parameters involved, and the fact that the dependence of the various directional stiffnesses on temperature is neither well studied nor readily available. Furthermore, the basic thermoelastic behaviour of composites has, until recently, not been understood. Even under negligible mean-strain situations, inconsistent results often arise from SPATE scans. This is due to the fact that the assumption of adiabatic conditions may not be valid in many applications. This was illustrated in Dunn *et al* (1989) in which the thermoelastic response of a particular composite specimen was shown to be highly frequency-dependent. Following this, a non-adiabatic theory for composite laminates was presented in Wong (1990) where it was shown that, in general, through-thickness heat diffusion cannot be neglected for laminated structures.

By considering a typical carbon-fibre/epoxy-matrix system, an order-of-magnitude analysis showed that under normal loading frequency conditions (i.e. 5–30 Hz) the fibre/matrix system may be treated as a lumped entity from a thermodynamic point of view. Within such a frequency range the individual constituents may be considered thermally 'thin', and that diffusion effectively homogenises the temperature between fibre and matrix over the timescale of each loading cycle. However, it was shown that homogenisation of the temperature between different laminae would require a much longer timescale. This means that different laminae may not be lumped in the analysis. On the other hand, they may not be treated independently either, as adiabatic conditions are achieved only at extremely high loading-frequencies (i.e. of the order of kilohertz).

By retaining the through-thickness heat diffusion term in the thermoelastic equation, i.e. taking

$$-Q_{i,i} = k_z \frac{\partial^2 T}{\partial z^2} \tag{7.38}$$

where $k_z$ is the coefficient of thermal conduction and $z$ is the through-thickness coordinate, it was shown that the frequency dependence of the thermoelastic response may be successfully predicted. Figure 7.8 shows the comparison of the experimentally measured and predicted thermoelastic response characteristics for an XAS-914C $\{(0°, +45°, -45°)_4\}_s$ carbon-fibre composite specimen.

The failure to achieve adiabatic conditions in composites in general has some interesting implications. First of all, the long-held belief that SPATE 'sees' only the stresses on the surface layer is immediately dispelled. Although only surface measurements are taken by SPATE, the fact that sub-surface stresses can influence the surface temperature response means that SPATE data would actually contain stress information from the lower plies. Furthermore, since it was shown that the contributions of the sub-surface stresses to the surface temperature are frequency-dependent, it is easily seen that, depending on the lay-up configuration, scans at different frequencies could yield quite different SPATE patterns. This makes the interpretation of SPATE scans of composites extremely difficult. Amidst this complexity, however, it was shown that, given a known lay-up configuration consisting of at least three different fibre orientations, and the knowledge of the thermal diffusive characteristics of the material, this phenomenon may in principle be exploited to yield individual strain components. The possibility of obtaining individual strain components without resorting to the solution of the

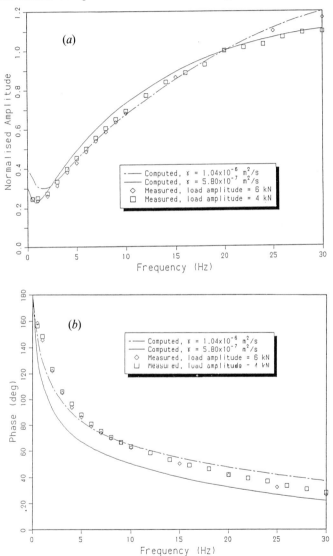

**Figure 7.8** Comparison of the measured and predicted frequency responses of the surface temperature of a carbon-fibre-reinforced plastic (CFRP) specimen: (*a*) amplitude response, and (*b*) phase response for two thermal diffusivity values.

equilibrium equation potentially makes SPATE a much more powerful stress analysis device for composites than it is for metals.

## 7.8 CONCLUSIONS

Although the development of the thermoelastic theory has been well established, recent interest in this phenomenon for the purpose of stress analysis and the development of an extremely sensitive dynamic temperature measurement device (SPATE) has led to the observation of anomalies which could not be explained by the existing theory. These were, namely, the mean-stress effect and, for the case of composites, the frequency dependence of the thermoelastic response. Both of these anomalies subsequently led to a revision of the current theory. In the case of the mean-stress effect, it was found that this phenomenon may be accounted for once the temperature-dependent stiffness terms are included. The physical basis for this is related to the fact that the process of thermal expansion is attributed both to the change in thermal excitation of the material and to the change in its stiffness.

The revised theory also predicted the existence of a second harmonic thermoelastic response when a single-frequency load is applied. Carefully executed experiments subsequently confirmed such a non-linear response. The discovery of this phenomenon also opened a new door for the development of a non-destructive means of detecting residual stresses within a material. The successful measurement of the residual stress profile of a four-point straightened specimen has already been achieved and a technique for tackling general two-dimensional residual stress problems is currently being studied.

For composites, the thermoelastic process is further complicated by inter-laminar thermal diffusion. A revised theory including the through-thickness heat conduction term was successful in explaining the frequency-dependent characteristics of the thermoelastic response. It was shown that whilst this effect makes the interpretation of SPATE results extremely difficult, it can in principle be exploited for the determination of individual strain components, thereby making SPATE potentially more useful for analysing composites.

## ACKNOWLEDGMENTS

The author wishes to thank R Jones, J G Sparrow, S A Dunn and R T Potter for their valuable input and comments concerning this work.

Figures 7.3 and 7.4 are reprinted by permission from *J. Phys. Chem. Solids* **49** 395–400 1988 (Oxford: Pergamon).

Figures 7.1 and 7.3 are reprinted by permission from *Nature* **332** 613–5 (copyright © 1988 Macmillan Magazines Ltd).

# 8

# The Thermoelastic Analysis of Anisotropic Materials

## R T Potter and L J Greaves

## 8.1 INTRODUCTION

Anisotropic materials are those whose physical and mechanical properties vary with direction. Wood is probably the most familiar example with its substantially different strengths along and across the grain. The majority of traditional engineering materials, however, do not exhibit anisotropy to any great degree and the development of theoretical and experimental methods of stress analysis has tended to concentrate on isotropic materials. In recent years the aerospace industry has pioneered the use of aligned fibre composites for load-bearing structures and this has created renewed interest in the stress analysis of materials with a high degree of anisotropy.

In this chapter we shall concentrate on orthotropic materials, that is materials whose property variations are symmetric about axes at right angles. The illustrations and examples will generally relate to carbon-fibre reinforced plastics (CFRPS) but the equations are applicable to all orthotropic materials and, with some modification, to more generally anisotropic materials. It should be noted, however, that materials with more general anisotropy are rarely used in practical load-bearing structures and components.

In describing the thermoelastic analysis of orthotropic materials we shall be concerned with several sets of orthogonal axes. Consider a multi-directional fibre-composite laminate which is subject to arbitrary in-plane loading, as illustrated in figure 8.1. The in-plane elastic properties of the laminate will vary symmetrically about the orthotropic axes, $x$ and $y$.

Thermoelastic analysis is based on the measurement of temperature fluctuations in the surface ply, whose in-plane orthotropic axes, 1 and 2, will not necessarily be coincident with those of the laminate $(x, y)$.

A third set of axes $(j, k)$ are those of the principal stresses in the surface ply and these will not generally be coincident either with the laminate orthotropic axes $(x, y)$ or with those of the surface ply $(1, 2)$. It should be noted, however, that failure mechanisms in highly anisotropic materials tend to be dominated by material anisotropy rather than by the orientation of the principal stresses or strains. For example, fibre composites generally fail due to the fracture or instability of the fibres or by the propagation of cracks running parallel to the fibres. In the design and evaluation of fibre-composite structures, the stress analyst is not therefore primarily interested in principal stresses and strains but in the stresses and strains related to the material coordinate system. The thermoelastic equations for these materials will therefore be expressed in terms of material axes rather than principal stress or strain axes.

It should also be noted that it is generally more convenient to work in terms of strain rather than stress in the analysis of laminated composite materials. The in-plane stresses within a particular ply depend upon ply orientation but are related to the laminate stress resultants by conditions of strain compatibility. Thus, we will here present the thermoelastic equations in terms of strains parallel to and at right angles to the fibres in the surface ply, that is, in the $(1, 2)$ axes. The subsequent calculation of stresses and strains in subsurface plies or in the laminate as a whole may be achieved using established methods of laminate analysis (see, for example, Ashton *et al* 1969).

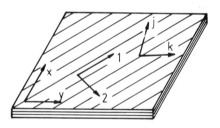

x,y   laminate axes
1,2   surface ply axes
j,k   principal stress axes

**Figure 8.1** Definition of axes.

## 8.2 THE THERMOELASTIC EQUATION

The general thermoelastic equation which relates the adiabatic change in temperature to a given change in strain was first derived for isotropic materials by Wong *et al* (1987). The more general form applicable to anisotropic materials has been derived (Potter and Greaves 1987) as

$$\rho C_\varepsilon \frac{\mathrm{d}T}{T} = \left[ \frac{\partial C_{ijkl}}{\partial T}(\varepsilon_{kl} - \alpha_{kl}(T - T_0)) \right. $$
$$\left. - C_{ijkl}\left( \alpha_{kl} + \frac{\partial \alpha_{kl}}{\partial T}(T - T_0) \right) \right] \mathrm{d}\varepsilon_{ij}. \tag{8.1}$$

In this equation the summation convention implies that, where a suffix is repeated in a product, the summation is made with the suffix increasing from 1 to 3. The symbols are explained in Chapter 7.

If a sinusoidally varying strain is applied, equation (8.1) can be integrated to give

$$\rho C_\varepsilon \ln \frac{T}{T_0} = \left[ \frac{\partial C_{ijkl}}{\partial T}\bar{\varepsilon}_{kl} - C_{ijkl}\alpha_{kl} \right] \Delta \varepsilon_{ijkl} \sin \omega t$$
$$- \frac{1}{4}\frac{\partial C_{ijkl}}{\partial T} \Delta \varepsilon_{kl} \Delta \varepsilon_{ij} \cos 2\omega t. \tag{8.2}$$

Note that, for an applied loading frequency $\omega$, the temperature variation has components both at $\omega$ and at $2\omega$. In practice, we generally measure the amplitude of the $\omega$-frequency component only, and this is given by

$$\rho C_\varepsilon' \frac{\Delta T}{T_0} = \left[ \frac{\partial C_{ijkl}}{\partial T}\bar{\varepsilon}_{kl} - C_{ijkl}\alpha_{kl} \right] \Delta \varepsilon_{ij}. \tag{8.3}$$

The above equations can be generalised even further to allow for the effects of absorbed moisture, which give rise to additional terms (Jones *et al* 1987). However, these terms are generally small and are here assumed to be negligible. A further simplification would arise if we could ignore the terms in $\partial C/\partial T$. The error introduced would depend both on the material properties and on the mean strain about which the specimen was cycled. (For carbon-fibre/epoxy-resin composites it has been estimated that errors of almost 10% could be generated for each $1000 \times 10^{-6}$ of mean strain.) However, if conditions are such that these terms can be neglected, equation (8.3) simplifies to

$$\rho C_\varepsilon \frac{\Delta T}{T_0} = - C_{ijkl}\alpha_{kl}\Delta \varepsilon_{ij}. \tag{8.4}$$

This is the linear form of the thermoelastic equation commonly assumed.

Except for totally anisotropic materials (such as triclinic single-crystals), the material symmetry will normally allow further simplification of this equation. In the case of isotropic materials, the diagonal terms of the thermal expansion tensor are equal and the other terms are zero. In addition the strain–stiffness products add up to the sum of the hydrostatic stress components. This is clearly also equal to the sum of the principal stresses. Among anisotropic materials, those most commonly used in engineering are orthotropic (aligned fibre composites, for example). In this case the diagonal terms in the thermal expansion tensor are not equal. They are, however, the only non-zero terms. In addition, there are no stiffness terms coupling dilatational stresses to shear strains ($C_{iikl} = C_{klii} = 0$ if $k \neq l$). The thermoelastic equation therefore becomes

$$\rho C_\varepsilon \frac{\Delta T}{T_0} = - C_{iikk} \alpha_{kk} \Delta \varepsilon_{ii} \tag{8.5a}$$

or (in full) (Riley 1974)

$$
\begin{aligned}
- \rho C_\varepsilon \frac{\Delta T}{T_0} = \; & (C_{1111}\alpha_{11} + C_{1122}\alpha_{22} + C_{1133}\alpha_{33})\Delta\varepsilon_{11} \\
& + (C_{2211}\alpha_{11} + C_{2222}\alpha_{22} + C_{2233}\alpha_{33})\Delta\varepsilon_{22} \\
& + (C_{3311}\alpha_{11} + C_{3322}\alpha_{22} + C_{3333}\alpha_{33})\Delta\varepsilon_{33}.
\end{aligned} \tag{8.5b}
$$

When using the SPATE method a free surface is being observed so plane stress conditions prevail. This means that $\varepsilon_{23} = \varepsilon_{13} = 0$ and $\Delta\varepsilon_{33}$ is a function of $\Delta\varepsilon_{11}$ and $\Delta\varepsilon_{22}$, that is

$$\Delta\varepsilon_{33} = -(C_{3311}\,\Delta\varepsilon_{11} + C_{3322}\,\Delta\varepsilon_{22})/C_{3333}.$$

Thus, although all three strain components must be included in equation (8.5), only two of them are independent.

If required, equation (8.5) can also be expressed in terms of stresses (in the material orthotropic axes). Because of the free surface, $\sigma_{33} = 0$ and the appropriate equation is

$$\rho C_\varepsilon \frac{\Delta T}{T_0} = -(\alpha_{11}\Delta\sigma_{11} + \alpha_{22}\Delta\sigma_{22}). \tag{8.6}$$

Note that, if the expansion coefficients were equal, this equation would reduce to a form similar to that which is commonly used for isotropic materials. If the term in $\partial C/\partial T$ may not be neglected, equation (8.5) would become

$$\rho C_\varepsilon \frac{\Delta T}{T_0} = \left[ \frac{\partial C_{iikk}}{\partial T}\bar{\varepsilon}_{kk} - C_{iikk}\alpha_{kk} \right]\Delta\varepsilon_{ii} + \frac{\partial C_{1212}}{\partial T}\bar{\varepsilon}_{12}\Delta\varepsilon_{12}. \tag{8.7}$$

This is the general form of the thermoelastic equation for orthotropic materials.

## 8.3 THE EFFECTS OF ANISOTROPY ON THERMOELASTIC ANALYSIS

The foregoing section presents the thermoelastic equations in a form most readily applicable to anisotropic, and in particular to orthotropic, materials. In this section we shall concentrate on the implications of anisotropy for the thermoelastic analysis of materials and structures. The terms involving $\partial C/\partial T$ will be ignored here since their effects are secondary and are not specific to anisotropic materials. In any case, they vanish when the mean strain is zero.

Of the wide range of materials used in structural applications, fibre reinforced plastics are the most anisotropic materials in common use. Indeed, the degree of anisotropy is such that structural components will often contain layers (or plies) of fibres in several different directions. In the following discussion it will be assumed that the temperature response is governed solely by the strains (or stresses) in the surface ply. The process is assumed to be wholly adiabatic and the stresses and strains in subsurface layers must be derived from those of the surface ply by laminate analysis.

Figure 8.2 shows the variation of adiabatic temperature with strain, as defined by equation (8.5), for a typical carbon-fibre/epoxy composite (CFC).

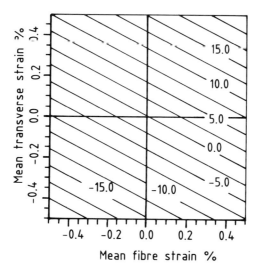

**Figure 8.2** Predicted SPATE output for a carbon-fibre composite ignoring temperature-dependent terms (arbitrary units).

In this figure the strain amplitudes are assumed equal to the mean strains in the fibre (1) and transverse (2) directions. The slanting parallel lines are contours of constant amplitude of the resulting temperature response (given in arbitrary units). The slope of these lines is a function of the material anisotropy and indicates the relative sensitivity of the technique to strains in the fibre and transverse directions. If the composite were equally sensitive to these strains the slope would be 45°, as for an isotropic material.

From equation (8.5a) it can be seen that the temperature change is related to each strain component via the coefficient $C_{iikk} \times \alpha_{kk}$. For a CFC, $C_{1111}$ is very large when compared to all the other stiffness terms but $\alpha_{11}$ is much smaller than $\alpha_{22}$. Thus, the coefficients relating to the fibre and transverse directions are of the same order of magnitude. This is fortunate since it means that, despite considerable material anisotropy, thermoelastic measurements (e.g. using the SPATE method) can detect high strains in any direction in a CFC. In practice, the temperature change due to transverse (2 direction) strains is about twice that due to fibre (1 direction) strains in these materials (Potter 1987) and therefore the contours in figure 8.2 have a slope of approximately $-0.5$.

The corresponding relationship between stresses and temperature can be seen from equation (8.6). If $\alpha_{11} \ll \alpha_{22}$, it might be tempting to neglect $\alpha_{11}\Delta\sigma_{11}$. However, because of the anisotropy of stiffness it usually turns out that $\Delta\sigma_{11} \gg \Delta\sigma_{22}$ and both stresses contribute significantly to the temperature variation.

In other fibre composites there is much less difference between the values of $\alpha$ such that, in some cases, the thermal response may be dominated by the strains (or stresses) in the fibre direction. Suppose, for example, a stress is applied to a unidirectional composite such that unit strain is produced in the loading direction. Then $\Delta\sigma = E$. The temperature change (from equation (8.6)) will be proportional to $\Delta\sigma\alpha$ and hence proportional to $\alpha E$. Comparing $\alpha E$ in the 1 and 2 directions will give an assessment of the importance of strains in the two directions for any given material. Typical values for a range of fibre reinforced plastics are presented in table 8.1. These data confirm that, for a CFC, thermoelastic measurements will be more sensitive to transverse

**Table 8.1** Comparative values of $\alpha$ and $E$.

| Material | Carbon/ epoxy | E-Glass/ epoxy | Kevlar/ epoxy | Boron/ epoxy |
|---|---|---|---|---|
| $E_{11}$ (GPa) | 130 | 40 | 75 | 200 |
| $E_{22}$ (GPa) | 9 | 8 | 6 | 15 |
| $\alpha_{11}$ ($\times 10^6$) | 0.43 | 6 | 4 | 18 |
| $\alpha_{22}$ ($\times 10^6$) | 28 | 35 | 40 | 40 |
| $\alpha_{11}E_{11}/\alpha_{22}E_{22}$ | 0.22 | 0.86 | 1.25 | 6.0 |

strains than to those in the fibre direction. For a boron/epoxy composite, however, one would expect thermoelastic measurements to be dominated by the fibre strains.

It is evident from the paragraphs above that the thermoelastic analysis of anisotropic materials should yield useful information provided the relationships between strains and thermal response are understood. However, for any material there are certain conditions under which finite strains will lead to a zero thermal response. For an isotropic material in a state of pure shear, there is always zero thermal response because the principal stresses and strains are equal in magnitude but opposite in sign. The conditions under which zero thermal response is generated in an orthotropic material are discussed below.

Consider first the case of pure shear. In applying the thermoelastic equation (8.5) to an orthotropic material we do not use the principal strains, but transform them into the material axes. Since this could involve a rotation through any arbitrary angle, the strains $\Delta\varepsilon_{11}$ and $\Delta\varepsilon_{22}$ will take values that, in general, will not lead to zero thermal response. For example, if the principal strain and material axes are coincident it can be seen from equation (8.5) that the thermal response will not be zero. There is, however, one case in which pure shear in an orthotropic material does give zero thermal response. This is when pure shear occurs in the material orthotropic axes (i.e. $\Delta\varepsilon_{11} = \Delta\varepsilon_{22} = 0$; $\Delta\varepsilon_{12} \neq 0$). This is the situation at the origin of figure 8.2.

The above observations might be considered an advantage since we have the prospect of detecting states of pure shear which would not be detectable in isotropic materials; only when pure shear occurs in the material axes will there be zero output.

Unfortunately, there is another situation in which no thermal response is generated whatever the strain amplitude. From equation (8.5), it can be shown that $\Delta T = 0$ when

$$\Delta\varepsilon_{22}/\Delta\varepsilon_{11} = \frac{-(C_{11kk}\alpha_{kk} + C_{33kk}\alpha_{kk}\Delta\varepsilon_{33}/\Delta\varepsilon_{11})}{C_{22kk}\alpha_{kk}}. \tag{8.8}$$

Since $\Delta\varepsilon_{33}$ is a function of $\Delta\varepsilon_{11}$ and $\Delta\varepsilon_{22}$, equation (8.8) defines a critical ratio $-v_0$ such that $\Delta T = 0$ when $\Delta\varepsilon_{22}/\Delta\varepsilon_{11} = -v_0$. (Alternatively equation (8.6) can be used to obtain $\Delta T = 0$ when $\Delta\sigma_{22}/\Delta\sigma_{11} = -(\alpha_{11}/\alpha_{22})$.) Thus, there is no thermal response whenever the ratio of strain amplitudes has the value $-v_0$. This condition is represented in figure 8.2 by the $\Delta T = 0$ line which passes through the origin and has a slope of $-v_0$. This situation arises, for example, in a uniaxially loaded specimen with a Poisson's ratio of $v_0$ and with the 1 direction of the surface ply lying parallel to the load direction. It turns out that many practical structural laminates made from CFCs have a Poisson's ratio fairly close to $v_0$. Fortunately, it is relatively uncommon for the outer ply to be parallel to the major load-axis otherwise such laminates would give only a small thermoelastic response in uniaxial tests.

In summary, the thermoelastic response of orthotropic materials is influenced by dilatational strains along the material axes, but by different amounts for the different directions. Shear strains referred to those axes make no contribution to the response. Zero response will be generated if the ratio between $\Delta\varepsilon_{11}$ and $\Delta\varepsilon_{22}$ takes a critical value or if both are zero. Note that it is the material axes that are of primary concern and, if these do not coincide with the structural or loading symmetry axes, asymmetrical patterns of response may be observed.

## 8.4 APPLICATION TO COMPOSITE LAMINATES

Having considered the theoretical aspects, let us now turn our attention to practical matters. The following observations relate to laminated materials made from plies of unidirectional fibre in a polymeric matrix (primarily carbon-fibre/epoxy). The thermoelastic response is monitored with a SPATE system, which measures temperatures by detecting infra-red emissions. Various factors must be considered if quantitative work is to be carried out and interpreted with confidence.

### 8.4.1 The effects of thermal conduction

The theoretical discussions in the previous sections have been based on the assumption of material homogeneity within the ply whereas, in reality, the microstructure consists of discrete fibres and matrix. For a given applied load amplitude, the heat generated in the fibres and matrix will give rise to different temperature changes. If the process were adiabatic, the observed temperature variations would be based on the average radiation emitted by the surface and this would be a function of the proportions of fibre and matrix exposed at the surface. However, the time taken to approach thermal equilibrium over distances of the order of the fibre radius (a few microns) can be shown to be about $10^{-5}$ s (Wong 1990). Thus, unless the loading frequency is very high, the process cannot be considered adiabatic. The fibres and matrix will normally be at the same temperature and this will be the temperature observed in thermoelastic measurements. Such measurements can therefore be interpreted using the mechanical and physical properties of the ply and it is not necessary to consider those of the fibres and matrix separately.

Whilst the heat transfer between fibres and matrix may be considered advantageous, departures from purely adiabatic conditions on a more macroscopic scale might pose more of a problem. The significance of these effects has been studied by thermal modelling (Wong 1990) and it has been shown that the heat loss to the environment is insignificant above about 1 Hz. Indeed, if the measurements were being made with a SPATE system, the signal amplitude at low frequency would be more severely attenuated by the circuitry used to eliminate the DC component than by the effects of heat loss.

The most significant of the thermal conduction effects arises from conduction between plies. For a laminate under uniform in-plane strain, the temperature generated within each ply will be a function of the fibre orientation. The combination of ply thickness, which is typically about 0.125 mm, and thermal conductivity means that the observed surface temperature may be significantly affected by thermal conduction between the surface and subsurface plies at loading frequencies below about 30 Hz. Typical graphs of SPATE output against frequency for an aluminium specimen, in which no such conductivity effects occur, and a multi-directional composite laminate are illustrated in figure 8.3. The curves have been normalised by plotting the SPATE output as a fraction of the output at 20 Hz. Wong (1990) has shown that the different shape of the two curves is entirely due to thermal conductivity between the surface and subsurface plies in the composite laminate. This phenomenon will cause particular problems in the interpretation of data from structural components in which the stress and strain state varies from point to point since the extent to which the surface temperature distribution (and hence the SPATE scan) is influenced by subsurface plies will depend both on the local stress state and the loading frequency. A related phenomenon may occur due to the presence of a surface layer of resin, as has been observed by Potter and Greaves (1987). However, the

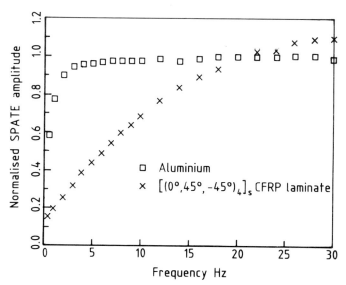

**Figure 8.3** Effect of loading frequency on SPATE output.

surface layer resulting from typical autoclave or press-moulding processes is generally very thin and the effect will not normally be significant.

Although thermal conduction effects appear to complicate the interpretation of thermoelastic data, Wong (1990) has proposed that, for multi-directional composites, the difference in response at three selected frequencies could be used to calculate the separate strain components.

A further point to be noted is that in the fatigue testing of fibre reinforced plastics various irreversible processes can cause the specimen to heat up if frequencies greater than about 5 to 10 Hz are used, particularly when associated with large strain amplitudes. This clearly has implications for the quantitative interpretation of thermoelastic data measured at quite low frequency but it may be particularly important in the observation of structural vibration modes in fibre-composite structures. Quantitative measurements of higher frequency modes may be strongly affected by the varying temperature of the structure. However, qualitative pictures will be enhanced because the more highly stressed areas, which are warmer, will give disproportionately high signals.

### 8.4.2 Material property data

In order to correlate theory with experiment it is necessary to know all the material properties appearing in the equations. These include thermal expansion and stiffness properties in all three directions. However, the scarcity of accurate material property data will often prove to be a problem for composites, especially where the properties in the through-thickness (3) direction are concerned. Indeed, in many instances there may be no data at all. Therefore, if a thermoelastic equation in terms of strain is to be used, some of the properties will have to be either measured or estimated. To some extent this problem can be circumvented by noting the relationship between the in-plane stresses and strains (the in-plane stiffnesses being generally well known) and using a thermoelastic equation based on stresses. However, the need for an accurate value of the thermal expansion coefficient in the fibre direction $\alpha_{11}$ can still be a problem; it is a small quantity highly sensitive to variations in the volume fraction of fibres in the composite. Values quoted in the literature, even for nominally identical materials, often differ widely. For this reason it is best to carry out an experimental calibration to find the value of $\alpha_{11}$ appropriate to the thermoelastic measurements (see, for example, Potter (1987)).

### 8.4.3 The effect of the temperature dependence of material properties

In section 8.3, the terms in the thermoelastic equation containing $\partial C/\partial T$ were ignored because they cause theoretical complications which are unnecessary in an illustration of the effects of anisotropy. However, these terms cannot in general be ignored as they may cause significant errors when using thermoelastic response to derive quantitative data for anisotropic materials. The presence of the terms results in the temperature response being a function not only of the oscillating strain amplitude, but also of the mean strain about which those oscillations occur. This phenomenon has been called the 'mean-strain effect' (or, in metals, the 'mean-stress effect'). Attempts to assess theoretically the significance of this effect in carbon-fibre composites have been hampered by the lack of accurate material property data. However, using assumed data, the predicted error in the SPATE output shown in figure 8.2 which arises from neglecting the effect is illustrated in figure 8.4. This latter figure shows the contribution of the $\partial C/\partial T$ terms, which must be added to figure 8.2 to obtain the total response. It may be seen that the effect is least along a line through the origin with a slope of about $-0.3$. This is perhaps fortunate since most structural laminates have a Poisson's ratio in the range 0.3 to 0.6 although, as stated earlier, this also corresponds to the situation in which the thermoelastic response is low. The errors are predicted to be quite significant when $\varepsilon_{11}$ and $\varepsilon_{22}$ have the same sign. It should be noted that these predictions are based on estimated material properties and require experimental confirmation. The available experimental results (see, for example, Dunn *et al* 1989) indicate an effect varying from zero to as much as 18% per 1000 microstrain depending on ply angles. Ideally, an experimental calibration of the thermoelastic response of a particular material should include some attempt to quantify the mean-strain effect for that material.

Apart from the effect mentioned above, the thermoelastic response will vary with the mean-strain if the material is not truly linear-elastic. Although

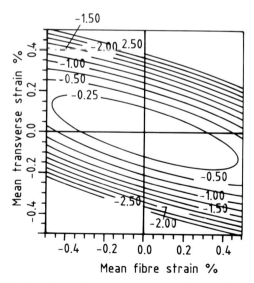

**Figure 8.4** Predicted error in the SPATE output due to the neglect of the temperature dependence of CFC properties.

most of the high-performance structural fibres are linear-elastic, most polymeric matrices exhibit non-linear behaviour to some degree. Thus, for laminates and loading modes in which the matrix is significantly stressed, the stress–strain relationship may be noticeably non-linear, that is, the stiffness may be a function of strain. Clearly, if equation (8.5) is used to quantify thermoelastic measurements in terms of strain then the results will be affected. In principle, this effect could be taken into account in calculations if the dependence of the stiffness on strain was known. However, since any of the stiffness terms could depend on any or all of the strains, this may be very difficult in practice.

It was noted above that consideration of the effects of moisture content gives rise to additional terms in the thermoelastic equation. While these terms themselves are negligible, the stiffnesses, $C_{ijkl}$, and to some extent the density are also functions of moisture content. For this reason it is possible that the thermoelastic response of certain composites (or, indeed, any hydrophilic material) could be significantly affected by absorbed moisture. Since the material that contributes to the observable response is close to the surface it does not take long to equilibrate with ambient conditions. This could result in day to day changes in the thermoelastic response according to the weather. As yet, no attempt to quantify this effect either theoretically or experimentally has been published and the significance of this problem remains unknown.

## 8.5 SOME EXAMPLES OF PRACTICAL APPLICATIONS

It is clear from the foregoing sections that the interpretation of thermoelastic data from composite materials is more difficult than for isotropic materials. The lack of material property data means that experimental calibration will generally be necessary, perhaps using a conveniently located strain-gauge rosette. Even then the interpretation of SPATE scans will require a sound understanding of the principles of thermoelasticity, some knowledge of the material properties and laminate construction, and an appreciation of the phenomena discussed above.

The following examples have been selected to illustrate some of the phenomena discussed in the foregoing paragraphs which are specific to the thermoelastic analysis of anisotropic materials.

Figure 8.5 (Plate 4) shows a scan of a carbon-fibre cloth composite subjected to a uniaxial tensile load parallel to the warp fibres. Despite the fact that the surface-strain field is virtually uniform, the weave pattern shows up quite clearly. The SPATE output from areas in which the weft fibres are at the surface is substantially greater than that from the areas where the warp fibres are at the surface. Indeed, it is possible to deduce from this scan that the cloth was a five-harness satin weave. As indicated in section 8.3, for a carbon-fibre reinforced plastic the thermoelastic response to strains transverse to the fibres is about twice that to strains parallel to the fibres and this agrees well with the scan shown in figure 8.5.

The second example, which is illustrated in figure 8.6, shows a scan of a buckled carbon-fibre composite panel. This panel was made using plies of unidirectional preimpregnate so that no weave pattern is seen. The panel was subjected to compressive end-loading whilst out-of-plane movement was restrained along all edges by the test fixture. The panel buckled into two half-waves along the length and a single half-wave across the width. It may be seen that, although the deflected shape and hence the in-plane surface strains of the buckled panel were symmetric about the panel axes, the SPATE scan is asymmetric. This is due to the fact that the surface ply lay at 45° so that the elastic constants and the expansion coefficients were asymmetric with respect to the panel axes. This panel would also have been subject to the thermal conduction effects discussed in section 8.4.1. Tests at different frequencies would therefore produce different asymmetric patterns. Note that the use of $+/-45°$ plies at the surface of structural laminates is common practice in aerospace applications for reasons of damage tolerance.

This latter example graphically illustrates the importance of knowing the laminate construction when interpreting SPATE scans. In the previous example the weave pattern makes it unlikely that the regular variations in output

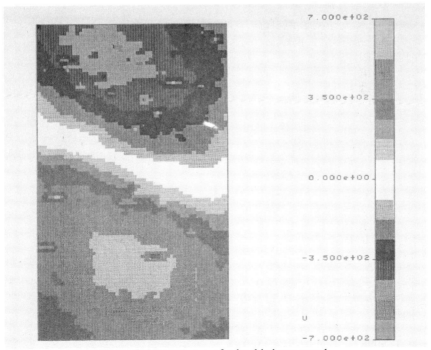

**Figure 8.6** SPATE scan of a buckled CFC panel.

would be misinterpreted. But in this case there are no such clues and the scan might well be misinterpreted as asymmetric buckling such as often occurs with anistropic materials. In the absence of other information, only a full quantitative analysis of the SPATE data, including conductivity effects, would reveal the symmetry of the buckle pattern with respect to the panel. In practice, the test engineer will often be aware of the elastic and loading symmetries and should not be misled by the asymmetry of the scan.

Figure 8.7 shows another carbon-fibre composite panel subject to buckling. This panel, however, has a thinner rectangular section in the middle, the transition between thinner and thicker sections occurring over a short taper. Again the surface ply is at 45° to the panel axes resulting in an overall asymmetry in the thermoelastic output. However, in addition to the variations in surface strain due to buckling there are strain concentrations at the four taper intersections. The initial impression gained from the SPATE scan is that these strain concentrations differ in magnitude. But, again, the difference is primarily due to the asymmetry of the elastic and thermoelastic properties of the surface ply. Note also that the apparent difference would change with frequency due to thermal conductivity from subsurface plies as discussed in section 8.4.1.

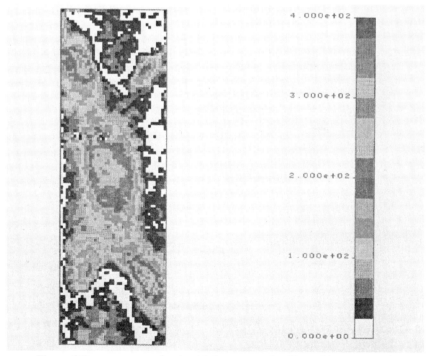

**Figure 8.7** SPATE scan of a buckled CFC panel with a thinner central section.

The foregoing examples have highlighted some of the problems associated with the thermoelastic analysis of anisotropic materials. The following examples illustrate two major benefits.

A particular problem with anisotropic laminated materials can arise from the interlaminar direct and shear stresses which occur at a free edge of a laminate under in-plane loading and which may initiate delaminations. Figure 8.8 shows two SPATE scans of the surface of a coupon specimen subjected to tensile fatigue loading. The first scan was generated early in the fatigue life and shows the stress concentration at the ends due to a reduction in thickness. After prolonged fatigue loading had given rise to edge delamination, the surface plies began to lift giving rise to a change in lateral strain which may be clearly seen from the second scan (figure 8.8(*b*)). The extent of the delamination as indicated by the SPATE scan has been shown to correspond accurately with that determined by ultrasonic measurements and the method has been successfully used to monitor the delamination growth during fatigue tests. The advantage of such a method is that it is non-contacting and does not necessitate the removal of the specimen from the test machine or even the cessation of the fatigue test.

Laminated fibre composites are also susceptible to low-velocity impact damage which tends to cause multiple delaminations, translaminar cracking and some fibre damage. Clearly, the measurement of the stresses or strains within the damaged region may only be done by a non-contact method such as thermoelastic analysis. Figure 8.9 (Plate 4) shows a SPATE scan of an impacted carbon-fibre composite laminate. The laminate was subjected to a small compressive fatigue load and, since the fibres in the surface ply were parallel to the surface, the SPATE output from areas remote from the damage was very low. At the geometric centre of the damage the surface fibres were broken and therefore there was no stress. On either side, the surface fibres remained intact although the ply was delaminated from the subsurface material and this allowed the ply to buckle out-of-plane. Whilst this buckling could have been detected by the naked eye, the SPATE scan shows that it was of sufficient magnitude to induce tensile strains at the surface. Perhaps more importantly, the scan also shows a significant compressive stress concentration at the sides of the impact damage due to a lateral redistribution of the compressive load. Such a stress concentration would be very difficult to detect and measure by other, more conventional means.

## 8.6 CONCLUDING COMMENTS

It is clear from the material presented in the present chapter that the use of thermoelastic analysis to evaluate stresses and strains in anisotropic materials is substantially more complex than for isotropic materials. The anisotropy of the mechanical properties and thermal expansion coefficients

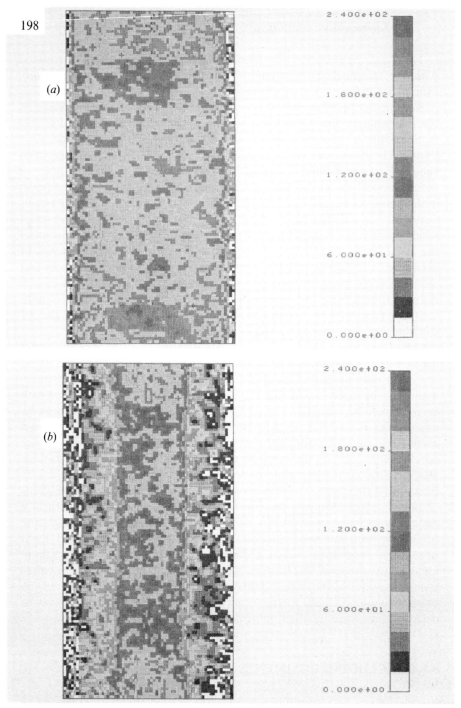

**Figure 8.8** (*a*) SPATE scan of a tensile fatigue specimen at the beginning of the test; and (*b*) SPATE scan of a tensile fatigue specimen showing the effect of edge delamination.

combine to make the subjective assessment of SPATE scans unreliable in the absence of adequate information on material properties and laminate construction.

A particularly useful feature of the method is the ability to make an area scan of a structure to detect unexpected stress concentrations. However, for any material there are conditions under which the temperature change will be zero whatever the stress or strain amplitude and it is therefore possible that the areas giving the highest change will not necessarily be the only areas of significance in terms of strength and durability of a structure. For an isotropic material, a zero temperature change occurs only for states of pure shear whereas, for an anisotropic material, zero change occurs both for pure shear in the material axes and for a particular ratio of longitudinal to transverse strain (or stress) which depends on the particular properties of the material.

In composites such as boron reinforced plastics the thermal expansion coefficients in the fibre and transverse directions are similar. The thermo-elastic response therefore tends to be dominated by the generally larger stresses parallel to the fibres although, in practice, failure may well be precipitated by the smaller, transverse stresses. In carbon-fibre composites the anisotropy of the expansion coefficients tends to counteract that of the stiffnesses and the response to typical fibre and transverse stresses (or strains) is of the same order. However, the response is rarely equal and care must always be taken in the interpretation of scans.

There are a variety of practical problems in deriving quantitative data, particularly for fibre reinforced plastics. These include inadequate knowledge of material properties and their variation with temperature, local variations in surface properties and the effects of absorbed moisture. However, these may largely be overcome by direct calibration using strain gauges. The effects of thermal conduction between subsurface and surface plies may also be overcome to a large extent by testing at high frequency, although this may not always be practicable, particularly for larger structures. Moreover, care must be taken to avoid significant heating of the material.

Despite the theoretical complexities and the practical difficulties, thermo-elastic methods have been used successfully on highly anisotropic materials, giving information which would be difficult, if not impossible, to obtain by other current techniques. With adequate experience of both the thermoelastic method and the behaviour of anisotropic materials it is possible to detect anomalies in the stress or strain patterns of complex structural components. Moreover, being a non-contact method it is possible to evaluate stresses and strains in and around areas exhibiting the complex damage formations typical of fibre-composite materials. Thus, particularly when used in con-junction with other methods, the thermoelastic analysis of anisotropic materials is likely to prove valuable in many areas, including research, product development and perhaps even as an inspection tool.

**ACKNOWLEDGMENT**

This chapter is Crown copyright.

# 9

# Post-processing of SPATE Data
## J T Boyle

## 9.1 INTRODUCTION

While progress and recent developments in the use of the SPATE system have been dealt with in preceding chapters, the aim here is to examine one of the more troublesome aspects of its practical use: namely, the problem of interpretation of the SPATE data in a form which is more meaningful for component design and assessment. This is tackled here in the context of modern Mechanical Computer Aided Engineering (MCAE) techniques. This chapter is aimed at novices in experimental thermoelastic analysis, like the writer, and at practitioners alike. While much of the discussion may seem rather obvious to the latter group, many significant arguments do not seem to have been stated clearly in the literature before. This gives the writer an excuse to rehearse them here.

Specifically, the problem of stress separation will be addressed: that is, at first sight, SPATE only provides a simple scalar quantity—the change in bulk stress (or hydrostatic (mean) stress; the first stress invariant). While this may be of use in some applications (a topic which will be discussed in more detail later on in this chapter), the immediate question is whether or not additional information, of more general meaning to the engineer, can be extracted from the basic SPATE data. The simple answer to this question is in the affirmative: by its nature SPATE is a full-field method of experimental stress analysis, and further it may be used repeatedly to 'zoom-in' on an area of interest. Thus, not only does SPATE provide the bulk stress, but also sufficiently detailed full-field data to derive significant information on the bulk stress gradient, and further the ability to interactively examine a localised area of interest in order to refine this derived data. If SPATE could only provide the bulk stress at fixed discrete sampling points, then there would probably be no wisdom in processing the data further; but it is important to realise that SPATE does provide more information than at first sight. The next question is, how do we achieve this? The answer here is not so simple, since it would depend on a clear understanding of two aspects of the use of SPATE which are not obvious; it is important to begin with a discussion of these aspects before proceeding.


## 9.2 FACING UP TO REALITY

Recall at all times the question which we are trying to answer: how can we extract further, more appropriate information from SPATE data? Before we can attempt an answer here, two features of SPATE need to be considered.

As discussed elsewhere in this book, SPATE is based upon the measurement of infra-red radiation which is emitted from the surface of a component under strain. The radiation is a result of a temperature change due to straining, a consequence of the generalised laws of thermodynamics for deforming bodies. For many engineering applications, this temperature change is small and can be ignored, but the pioneering study of Belgen (1968) demonstrated that modern IR radiometric techniques could be used to measure this temperature change.

So far so good! The first problem lies in the interpretation of this signal, and it is here where some significant assumptions are introduced. In order to interpret the temperature change in terms of mechanical quantities it is necessary to make some assumption about the material properties. Thus in the first instance it is assumed that the material is isotropic, homogeneous and linear-elastic (the last of course also being related to the load level on the component under consideration). With these assumptions, the thermoelastic theory due to Lord Kelvin (Thomson 1853) and Biot (1956) can be used to show that the SPATE signal is proportional to the bulk stress. If these assumptions are removed then the interpretation of the SPATE signal becomes more troublesome, although some progress can be made for certain classes of composite materials (see Chapter 8) and in the thermoplasticity of metals.

If the material is not homogeneous (as in a welded, cracked or damaged component, or locally in a composite with finite structure) then great care must be taken in the interpretation of the SPATE thermal signal since an adequate thermodynamic theory of non-homogeneous materials is not available, and it is necessary to show that the engineering practice of homogenisation of the mechanical variables and associated field equations is applicable locally to the thermodynamic equations for rapid cyclic-loading, questions which have not yet been addressed in the context of SPATE.

Nevertheless, within the confines of these basic assumptions, which do apply to the majority of engineering design situations, SPATE has been shown to perform very well! and it is in the application to these problems that we will continue. The point to be remembered is that, even in the simplest case, in order to interpret the SPATE signal we have made assumptions about the material behaviour! A clear understanding of the reality of this assumption is necessary to answer the question of how to process further the SPATE data.

The second aspect relates to a more fundamental question: why is SPATE being used in a specific experimental situation? A design and development engineer who is interested in a real component wants to know how his component behaves under design and operational loads and to make sure that his design assumptions are correct and that the component is not over-stressed. It is hoped that he would be using SPATE as one tool in his assessment.

The writer, who is not an experimentalist, can take the opportunity to be facetious here: the aim is not just to be able to accurately measure the 'actual stress' in a component (which is often an illusion), but is much more complex. To understand this an example can be given: the following quote is deliberately taken out of context from an unreferenced, but real, source: '... it may be shown that SPATE would not be able to distinguish an unloaded body and one which is subjected to pure shear. The *failure* of SPATE to respond to pure shear was clearly shown ...' (the writer's italics). This statement is fair enough, SPATE would not be able to tell the difference since the bulk stress is zero in both cases: but SPATE is not intelligent and thus cannot use any additional information which is available—the engineer knows the loading applied, and with this additional information the SPATE data is not a failure, but a dramatic confirmation of an expected stress distribution in part of a component under pure shear! The point is made that the design engineer, or analyst, or plant engineer or whoever would not blindly point SPATE at a component to establish the 'actual stresses' and be done with it (it is hoped, but not of course always the case).

Two further examples are significant here: the design engineer may have finite element models of a range of expensive components which he wishes to verify by testing on a typical component under a specific loading condition; the forensic engineer may have a failure whose cause he needs to identify, and he suspects may be traced to an unexpected over-stress. The use of SPATE in the first example is fairly clear: the FE model and analysis may be used to provide visual contours of bulk stress which may be compared to the SPATE results. Regions of possible stress concentration, identified by the FE analysis, could then be further examined by a closer inspection with SPATE. Both techniques give full-field solutions which can be compared and fine-tuned to verify the FE model. The use of SPATE in the second example is not so straightforward since the question which must be answered is not clear. SPATE can be immediately used by simply pointing and shooting at an actual component in service, which has not failed, to assess the possible stress level and distribution under service conditions. This is the strength of the technique (although as the reader will be aware, it is not quite as simple as this due to the nature of the SPATE test procedure). This result may then be used, for example in a residual-life calculation to assess the continued service of this specific component. But, suppose the stress levels found do not reflect the initial design stresses; this could be due to bad or clumsy design, but it could also be due to unexpected loading on the component, or manufacturing problems, perhaps resulting in weak regions. The important point here is that SPATE could be used to address these problems also: given the measured bulk stress in regions where the SPATE material assumptions are valid, can the inverse problem be solved to determine the load condition or the material stiffness?

The examples given here are important since they reflect on the question of how the full-field SPATE data should be further processed. In the case of the verification of the FE models, there is probably no need for stress separation, since the FE results can be given in terms of the scalar SPATE bulk stress. What is important here is that SPATE should be used interactively to further access regions of interest. In the case of the failure assessment, several distinct procedures are possible. For a remnant-life calculation on an in-service component it would be necessary to perform stress separation for a fatigue or damage calculation. For the assessment of the reality of the design conditions (loading and material) a much more complex type of analysis (for the inverse problem) is more appropriate.

The point to be remembered here is that the specific MCAE technique which needs to be adopted depends on how and why SPATE is being used, a simple development of stress separation techniques alone may not be either appropriate or necessary. Further processing of SPATE data depends on an understanding of this reality also.

## 9.3 POST-PROCESSING SPATE DATA FOR BIAXIAL PROBLEMS

We begin here with a study of the simplest case which may arise: the processing of SPATE data from biaxial problems—planar components wholly under loading in the plane, typically reflecting mechanical plane stress problems. While such problems do arise in engineering applications, they are mainly used here as test problems for SPATE; indeed many applications of SPATE to this class of problems can be found in the literature.

Here we will discuss the possibilities for further processing of SPATE data, including the direct problem of stress separation. This discussion will be used as a basis for examining the more general problem of real curved three-dimensional structures; but, before proceeding, a basic difference between the two- and three-dimensional problems needs to be highlighted. In the biaxial case the boundary data is usually exact (such as a free boundary, or a symmetry line) or to be determined (for example in the inverse problem) with the SPATE data supplying interior data. In the three-dimensional problem, SPATE provides detailed surface boundary data in addition to known conditions, but not interior data. In the following we will discuss three possible ways of processing and using SPATE data for biaxial problems. Only one of these is relevant in the three-dimensional problem! Thus, while it is tempting to get over-enthusiastic for ingenious analytic methods for stress separation in the biaxial case, it should be remembered at all times that this will probably not be applicable to the three-dimensional problem.

The problem of stress separation has been reviewed previously in several publications (Rowlands 1986, Stanley 1986, Huang *et al* 1988) and it is fair to say, without any intended criticism, that no generally applicable technique has been demonstrated for biaxial stress separation from real SPATE data.

Indeed the writer has found considerable confusion here which this chapter aims to clarify as well as placing in perspective. The basic problem of stress separation from SPATE data is not difficult for biaxial problems and forms a fairly simple boundary value problem, if the SPATE data is concurrently smoothed. Difficulty may arise if this is not how SPATE is being used for a particular application—for example, applied to an inverse problem of determining boundary data or material properties.

In the following, three distinct approaches to the post-processing of SPATE data will be discussed. Firstly, the problem of stress separation from stress equilibrium considerations alone will be examined. It is shown that this gives rise to a simple, but uncommon, boundary value problem which may only be resolved in the general case of a component with arbitrarily curved boundaries through the development of new algorithms. Secondly, it is argued that the nature of the basic inherent SPATE assumptions, in particular the specific material properties which are presumed, constrain this problem unnecessarily (since new analysis techniques need to be developed and tested) and since smoothing is essential. If the material assumption is also admitted in the stress separation problem, then it is argued that existing and familiar analysis techniques may be used. Finally, approaches for the inverse problem are examined. The confusion which has been apparent in the literature is directly related to an apparent lack of a perceived difference between the analytical requirements of these three approaches.

### 9.3.1 Equilibrium methods of stress separation

At first sight the problem seems quite simple: given the bulk stress over the interior of a two-dimensional solid, and known boundary conditions, is it possible to extract the stress tensor at sampling points? Such stress separation is a well-known problem in other full-field experimental techniques such as photoelasticity, moiré interferometry and image analysis, but is less of a problem since these techniques provide more information. SPATE only provides scalar bulk stress, while these other techniques can also provide vectorial (directional) information, but is a full-field analysis with high resolution.

With the bulk stress alone, it is not possible to extract stress components algebraically, since we really only have the 'average' of the stress tensor. To do so it is necessary to examine the stress-field equations of equilibrium and apply some appropriate analytical technique which makes use of the full-field data to provide gradients of the average stress. Some theory first:

### 9.3.1.1 Basic equations for biaxial stress

Let $(x, y)$ be some suitable reference coordinate frame, as shown in figure 9.1. We consider some deforming component as shown.

For biaxial problems, the state of stress at a point is fully determined by the three distinct components of the symmetric stress tensor, $(\sigma_x, \sigma_y, \tau)$, being the normal stresses and shear stress on two orthogonal planes through the point parallel to the coordinate axes. The stress is assumed to vary continuously from point to point.

At this point it is worth being pedantic and remind the reader that the concept of a continually varying stress, and that of stress defined at a point, is a mathematical one relying on a limiting process for the normal and shear forces on a small area which collapses onto the point. This point is significant since it is possible to forget when 'measuring stress' experimentally. Recall, SPATE is measuring temperature change, and converting that to stress through the equations of continuum thermodynamics. For most engineering

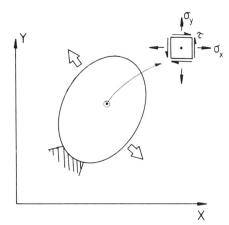

**Figure 9.1** Biaxial stresses.

purposes this is satisfactory, but, again, if the high resolution of SPATE is being used on a local area (such as a region of damaged material) due to a weld or fatigue micro-cracks, or a composite with finite structure, care must be taken since this definition of stress may not be applicable—indeed can only be used through the process of homogenisation (Christensen 1979), which may not always be successful.

If the stresses can be represented as continuously varying, then their first-derivatives exist, and the familiar equilibrium equations in rectangular coordinates may be derived (Fung 1965):

$$\frac{\partial \sigma_x}{\partial x} + \frac{\partial \tau}{\partial y} = 0 \qquad \frac{\partial \sigma_y}{\partial y} + \frac{\partial \tau}{\partial x} = 0 \qquad (9.1)$$

ignoring body forces (such as centrifugal forces or gravity).

Here we have two equations in three unknowns, and except for special statically determinate problems, these cannot be solved in general; they must be completed by the other field equations (strain–displacement relations, and the stress–strain law) to find a solution. However, with SPATE we also know the bulk (mean) stress

$$\sigma_m = \sigma_x + \sigma_y \qquad (9.2)$$

which is the first stress invariant, and thus also equal to the sum of the principal stresses. We may then eliminate either of the normal stresses to give either

$$\frac{\partial \sigma_x}{\partial x} + \frac{\partial \tau}{\partial y} = 0 \qquad -\frac{\partial \sigma_x}{\partial y} + \frac{\partial \tau}{\partial x} = -\frac{\partial \sigma_m}{\partial y} \qquad (9.3)$$

or

$$-\frac{\partial \sigma_y}{\partial x} + \frac{\partial \tau}{\partial y} = -\frac{\partial \sigma_m}{\partial x} \qquad \frac{\partial \sigma_y}{\partial y} + \frac{\partial \tau}{\partial x} = 0 \qquad (9.4)$$

which represent two equations in two unknowns in each case. Given suitable boundary conditions, these equations form two possible boundary value problems for one normal stress and the shear stress; for convenience we will call these equations the first-order form.

Before proceeding to a discussion of the boundary conditions, another form of these equations can be derived. It can be shown, by differentiating each set of equations given above with respect to $x$ or $y$ and then combining, that either set can be replaced by the equations

$$\nabla^2(\sigma_x - \sigma_y) = s_1 \qquad \nabla^2 \tau = s_2 \qquad (9.5)$$

where the functions $s_1(x, y)$ and $s_2(x, y)$ are given by

$$s_1(x, y) = \frac{\partial^2 \sigma_m}{\partial y^2} - \frac{\partial^2 \sigma_m}{\partial x^2} \qquad s_2(x, y) = \frac{\partial^2 \sigma_m}{\partial x \partial y}. \qquad (9.6)$$

These equations are immediately recognisable as Poisson equations, and are well known in the literature on partial differential equations; they are interesting for another reason as we will discover in a later section. We will call these equations the second-order form.

With any of these sets of equations, two factors serve to complicate matters:

(i) Firstly, it is necessary that the bulk stress, $\sigma_m$, is also differentiable. If one of the first-order forms is used, then only one first-derivative in either coordinate direction is required. If the second-order forms are used, then all second-derivatives are required. Thus the first-order forms, which only require one derivative, are to be preferred. But, as we will see later, there is a different reason for choosing the Poisson form, which turns out to be more convenient. In any case, the basic SPATE data must be smooth enough to determine its gradient (and possibly second-derivatives also).

(ii) Secondly, for a general curved boundary, the boundary conditions are given in terms of combinations of the normal and shear stresses. This is obvious for the first-order equations, since two first-order partial differential equations are used with two unknown functions (one normal and one shear stress). This is an inconvenience for the second-order Poisson equations, since, as we will see, solution methods for these are readily available if taken separately; but the nature of the boundary conditions on a general curved boundary prevents this, except in special cases where the boundaries are straight and parallel to the coordinate axes. The second-order equations are coupled.

Both of these problems will be discussed further below. However, it has been shown that the problem of stress separation, using equilibrium alone, reduces to the solution of differential field equations forming boundary value problems. This precludes, again except in special cases from straight boundaries (or rotationally symmetric using a polar coordinate system), any simple 'marching' technique from one free boundary to another (as can be done for example with 'graphical' integration methods applied to the equilibrium equations in stress separation for photoelasticity (Dally 1987b)). A complete boundary value problem must be solved! In theory this can be done, but there are two practical problems to be confronted.

### 9.3.1.2 The problem of smoothing

A cursory glance at typical SPATE contour plots illustrates two features: a certain amount of noise in the signal (which is to be expected), and a frequent 'edge error' at the boundary of biaxial benchmark problems. It is necessary then that the received SPATE data needs to be smoothed. In addition, for this specific application, the smoothing process would also be carried out in such a manner as to provide an analytic representation of the discrete data, which would allow differentiation of the latter.

In general, the smoothing techniques for experimental data are well established. Indeed two features of SPATE serve to work in our favour here.

Firstly it is only necessary to smooth a scalar quantity; much more care needs to be taken with full-field experimental data which represents vectorial data (Feng and Rowlands 1987), since small errors in vectorial data can give a significant loss in accuracy when differentiated.

Secondly, the SPATE data represents a real physical result which is expected to be smooth (and inherently differentiable), although again within the confines of the basic assumptions concerning the material behaviour (essentially isotropy and smoothness), and analytic surface fitting (interpolation) techniques (Lancaster 1986) should be able to be applied rather than statistical (regression) techniques (Brandt 1976). Smooth surface interpolation with bicubic splines, piecewise polynomials and more esoteric concepts, such as Kuhn's patches, are well documented, and much software is readily available (NAG 1987).

The effect of noise and general smoothing of the data is arguably less of a problem than that of edge error; in the case of SPATE it is probably best to use engineering judgment and ignore the edge values, relying on smooth extrapolation from close interior data. The effect of such procedures has not been examined in any detail. However this problem is not unknown in a related application. The inverse problem in heat conduction has become of interest recently, where uncertain boundary data is derived from measured interior data (Beck *et al* 1985) and, if it were thought sufficiently critical, such techniques could be used to derive more consistent SPATE boundary data from an analysis of the thermal signal itself. Other, and indeed related, inverse problems in post-processing SPATE data will be discussed below.

SPATE bulk-stress data is a sufficiently detailed representation of a real smooth scalar physical quantity to expect successful smoothing, and to allow for the confident evaluation of both first- and second-derivatives of the data. Any post-processing of the data would demand this (although, as we will see, some techniques may only require fairly crude smoothing).

### 9.3.1.3 The problem of boundary conditions

The only boundary conditions which we can use in a purely equilibrium method are the so-called natural boundary traction conditions. Fortunately this covers many applications for which SPATE is used, and in particular the type of biaxial benchmark problem seen in the literature.

With reference to figure 9.2 the boundary conditions are

$$\sigma_x n_x + \tau n_y = t_x \qquad \tau n_x + \sigma_y n_y = t_y \qquad (9.7)$$

where $(n_x, n_y)$ is the normal to the planar boundary, and $(t_x, t_y)$ the specified boundary tractions. The tractions would be zero for a free boundary, or specified at regions of load application. The loads may of course be given as a total value over part of the boundary rather than distributed as point tractions; well-known techniques for integrating these are available in the literature (Timoshenko and Goodier 1982).

For a solution of either the first-order or second-order form of the equations, these conditions would need to be given on the whole of the boundary; otherwise purely equilibrium techniques are impossible in general, and hybrid methods would be required.

Two conditions (representing the traction vector) are sufficient to solve this problem (a general mathematical analysis of the sufficiency of these conditions is not difficult, but the writer has not examined this in any detail). Immediately we can use either set of the first-order equations with these conditions, replacing either of the normal stresses using the relation

$$\sigma_m = \sigma_x + \sigma_y \tag{9.8}$$

but note that this then requires the use of boundary data for the bulk stress $\sigma_m$. The need for smoothing, as discussed above, is thus essential for this method to work.

Techniques for the solution of the first-order equations are plentiful. With sufficiently smoothed and differentiable bulk-stress data, the most general method for arbitrarily curved boundaries would be based on finite elements. This could probably be intelligently coupled with the FE smoothing technique of Feng and Rowlands referenced above; indeed Huang and Rowlands (Huang *et al* 1988) have demonstrated an ingenious equilibrium technique for the special case of tensile plates with holes or notches based on this. An even better possibility for biaxial problems would be the use of boundary element (BE) methods, since, as we will discuss for a simpler problem below, this could allow only selected random-sampling of the interior SPATE data, with robust extrapolated boundary data. The reader is invited to be the first to try this.

Thus as stated in the preceding, the problem of stress separation can be represented by a simple coupled boundary value problem for one of the

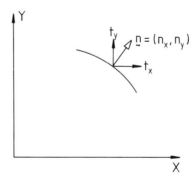

**Figure 9.2** Boundary conditions.

normal stresses and the shear stress, using the available, but smoothed, SPATE data, with the restriction of a class of biaxial benchmark problems with free or loaded boundaries. However, before the reader initiates the effort required to implement the above, he would be better advised to consider the implications of hybrid methods and the inverse problem, and the extension to the three-dimensional problem.

### 9.3.1.4 *The thermal analogy*

Before proceeding it is perhaps worth considering the implications of the use of the second-order form of the reduced equilibrium equations which give rise to Poisson equations. The immediate disadvantage is that second-derivatives of the SPATE data are required; although for the reasons given above this should not be viewed with too much concern. However, there is, for a class of benchmark problems, a noteworthy advantage in using the second-order Poisson form. This is the so-called thermal analogy:

> The Poisson equation, and indeed its simpler homogeneous form, Laplace's equation (Farlow 1982), can be found in diverse applications in science and engineering, for example describing torsion in a bar, the potential of an electrostatic field with constant charge density, the Helmholtz equation for vibration of a membrane, seepage and consolidation problems in soils, and, more importantly, two-dimensional steady state heat transfer with internal heat sources (an interesting analogy itself in the context of SPATE). Its solution has been studied in great detail; for arbitrary shapes the FE method is used almost without exception. The interesting feature from our point of view, is that if a steady state FE heat transfer analysis capability is available, then this can be used for stress separation by the thermal analogy! Commercial personal computer (PC) -based FE software (ANSYS 1987) in fact illustrate solutions to such analogous problems for the Poisson equation.

But the reader should not be too thrilled; the problem, for a general curved boundary, is that the boundary conditions for each of the Poisson equations in normal stress difference and shear stress respectively, require these equations to be coupled! That is, the boundary condition is not uniquely given in terms of either normal stress, or shear stress, but a combination of both. However, in the special case of components with straight edges, parallel to the rectangular reference frame, the Poisson equations can be uncoupled: either of the two traction conditions will give a suitable boundary condition for the unknown in the appropriate Poisson equation, and it is well known that this problem is well posed (that is, a unique solution exists). Thus, for this very restricted (but nonetheless still useful) class of benchmark problems, existing powerful commercial FE software is easily available. However, again interior smoothing for differentiation and for extrapolation to the boundary of the bulk stress is necessary. Results of this type of analysis for real smoothed SPATE will be published elsewhere.

On the other hand, the reader should not despair. The general problem for a curved boundary can be tackled, but not directly. Two analyses, one for each of the Poisson equations, with assumed uncoupled boundary data, can be solved, and then the results matched to the true coupled (traction) boundary conditions at the boundary nodes. Since the problem is linear, this matching will result in a set of simultaneous equations which can be solved for the actual uncoupled condition (that is the individual stress components are thus established). The analyses may then be repeated (or combined, being fairly straightforward in ANSYS) for the interior.

It was mentioned in the above that for the two-dimensional problems being considered here, the BE method could provide an interesting alternative (this is discussed later also for hybrid and inverse problems). With the current state of computer and software development, and in particular with the availability of solid modelling, automatic mesh generation and adaptive meshing (for example ANSYS Revision 4.4 includes post-error estimates with manual remeshing), the BE techniques offer no advantage over the FE method, since biaxial problems can be solved quickly on personal computers, and in seconds on state of the art engineering workstations (a situation, sadly, for the proponents of BE analysis as a replacement for FE techniques, which can be expected to get even better in terms of affordable desktop computing power in the near future). The role for BE analysis would appear to be in the application to problems where the FE method is not well suited (again, unfortunately not in the class of routine engineering stress or thermal analysis), for example near-field (crack) or far-field (semi-infinite media) problems. Interestingly, the stress separation problem can also arguably be better addressed using BE methods. The reason for this is not obvious, and requires some understanding of the BE method, and some interesting recent research. Here a simple explanation will be given in terms of the Poisson equation, by the thermal analogy.

Let us look at the Poisson equation

$$\nabla^2 \phi = s \tag{9.9}$$

where $\phi$ represents either the normal stress difference, or the shear stress, with suitable conditions on $\phi$ being given on the boundary (for the uncoupled problem for straight edges, of course recognising that a post-coupling procedure, as described above could be used for curved boundaries), and $s(x, y)$ is prescribed from second-derivatives of the bulk stress. The boundary integral method replaces this differential field equation by an equivalent boundary integral

$$c\phi_i + \int_\Omega s\phi_i^0 \, d\Omega + \int_\Gamma \phi \frac{\partial \phi_i^0}{\partial n} d\Gamma = \int_\Gamma \frac{\partial \phi}{\partial n} \phi_i^0 \, d\Gamma \tag{9.10}$$

where $\Omega$ is the region under consideration (figure 9.1), $\Gamma$ is the boundary of this region, the constant $c = \frac{1}{2}$ for a boundary point ($c = 1$ for an interior point for post-processing). The function

$$\phi_i^0 = \frac{1}{2\pi} \ln\left(\frac{1}{r_i}\right) \tag{9.11}$$

is known as the fundamental solution corresponding to some point $P_i$ in the region $\Omega$ and is necessary for the boundary integral formulation. For the homogeneous Laplace equation the area integral in $s(x, y)$ disappears, and the differential boundary value problem is replaced by an integral equation for the boundary only. Boundary element formulations may then be used to solve for the unknown normal derivatives of $\phi$ on the boundary (since in our problem it is assumed that the value of $\phi$ is given everywhere on the boundary). Only the geometry of the boundary needs to be discretised, not the whole interior as in the FE method. Unfortunately in our case we have a Poisson equation, so that the area integral must be evaluated (although only boundary elements are required once these integrals are known).

The usual method of calculating these area integrals has been to replace the interior with 'finite elements' for integration purposes only (there is no need to solve for $\phi$ at the interior nodes, only on the boundary). Many of the advantages of the BE procedure are lost in general. However another method has recently been suggested (Gipson 1989), where the integral is evaluated using Monte Carlo methods. This is a curious, but powerful, technique, based on probability theory and the theory of random walks to simulate physical processes, which may be described by differential and integral calculus. The simplest example in integral calculus is to use Monte Carlo techniques to evaluate integrals by random sampling of the integrand in the interior, and using the law of large numbers to approximate the integral. This is the technique which Gipson has suggested for evaluating the area integrals for the Poisson equation. In the past this random sampling method has not been efficient, due to the large number of calculations required. But with modern computers this restriction is removed. The attraction of the method in the stress separation problem for SPATE is that only a random sample of the interior of the SPATE data is required (possibly 1000–3000 points) and thus smoothing, for the bulk-stress differentiation, becomes a local problem at the sample points and at the boundaries only. An analysis of this possible technique will be published at a later date.

As a diversion, it may be interesting to note that Monte Carlo methods can also be used to solve partial differential equations very effectively in cases where boundaries are irregular, and results are only required at a few interior points (rather than everywhere). The boundary value problem does not need to be solved, but a large amount of sampling during so-called random walks through the interior is required. Such a technique could obviously be used at the time of SPATE testing to separate stresses at a few selected points, since no complex mathematics or FE or BE analysis is required, only sampling of the SPATE data! So how is this miracle achieved? The writer has not yet derived the necessary Monte Carlo simulation, which is the difficult bit, but invites the reader to attempt this. A good starting point is the drunkards walk in Lesson 43 of the book by Farlow (Farlow 1982), then the classic monograph by Hammersley and Handscomb (1964) and some inspiration!

### 9.3.2 Hybrid methods

It should be apparent to the reader that a fair amount of progress can be made with purely equilibrium techiques for stress separation. Nevertheless, the writer would question whether the effort is entirely necessary. As we will see, these equilibrium techniques cannot be extended to the three-dimensional problem (for a very simple reason). Further, and perhaps equally important, one of the basic SPATE realities has been neglected. Using equilibrium techniques alone avoids any need to make assumptions about the material behaviour (since only the field equations of stress equilibrium are invoked). This would be a good approach if SPATE indeed measured directly the bulk stress—of course it does not do this, it measures IR radiation and converts this to bulk stress with the theory of continuum thermo-dynamics together with assumptions about the material behaviour! Equilibrium techniques would be the way to proceed if the material behaviour was unknown, but we have assumed an isotropic, homogeneous linear-elastic material in order to convert the SPATE signal (the picture on the screen) to a mechanical variable; bulk stress. In order to interpret the SPATE signal for other materials, such as orthotropic homogeneous (laminated composite) materials, similar assumptions on the constitutive relations must be made.

So, if these assumptions are made at the outset, why can we not use them for stress separation? As we will see, it makes life much easier, and brings us into the realm of the so-called hybrid methods.

Hybrid methods have received much attention among the experimental stress analysis community (Kobayashi 1987). It is fair to say that an exact 'definition' of a hybrid method in experimental stress analysis is difficult. To some authors, any post-processing of experimental data, as in the equilibrium stress separation described above which relies on the equilibrium field equations of solid mechanics, would be called a hybrid method. To others, the class of inverse problems, described in the next section, would also be described as hybrid. However great care is needed with inverse problems, as we will see, and the writer prefers to treat these separately. To avoid confusion, hybrid methods are classified here as experimental post-processing techniques which are combined with results of conventional, independent stress analysis in some way. Two different methods are used to examine the same problem, experimental and theoretical (replacing the use of a further experimental method which would have been the approach in the past), and to enhance their respective results. In the spirit of Kobayashi '... the outcome ... is an improved mathematical model which yields complementary results to those obtained experimentally ...'. Thus, the equilibrium stress separation methods would arguably not be classified as hybrid, since the associated theoretical (numerical) analysis is not an independent physical model. (At this stage the reader will appreciate that the distinction is really arbitrary and that the writer is being pedantic.) The experimental and theoretical stress analyses may be combined in different ways, as may be expected. For example, experimental boundary data have been used (Gilbert *et al* 1987) as input to a reduced finite element model of a region of high stress gradient, avoiding the need for a more complete FE model (in FE terms this could be called 'hybrid substructuring'; experimental results, rather than a crude FE model, are used as boundary data for the substructure). A similar technique could be difficult using SPATE which would not in itself provide enough 'boundary' data for the FE substructure. Others have adopted a different approach (Berghaus 1989): photoelastic solutions were used to generate displacement boundary conditions to replace uncertain applied force conditions (as an extension of previous work by Chambless *et al* 1986). A similar approach was used with laser speckle holography (Weathers *et al* 1985), and using boundary integral methods (Balas *et al* 1983, Moslehy and Ranson 1980). In these cases caution is to be advised, since this approach in effect involves an inverse problem!

Hybrid techniques have been used previously for the stress separation of measured bulk-stress data. Balas (1967) demonstrated an ingenious finite difference analysis of the isotropic elasticity equations which used bulk-stress data to advance a solution from a free boundary, without the need to solve a complete boundary value problem. Unfortunately, this technique requires the boundaries to be straight and orthogonal, and cannot be adapted (as far as the writer can see) to arbitrary curved boundaries. Huan *et al* (1988) showed how classical elasticity solutions based on the Airy stress functions for polar coordinates could similarly be used, combined with a least-squares approach. In fact there is a fair amount of interest in using the classical elasticity solution techniques for hybrid methods in photoelasticity (Thompson *et al* 1975, Cavaco and de Freire 1989). These are based on the so-called 'methods of stress separation based on compatibility' where photoelasticity supplies the principal stress difference, and thermoelastic stress analysis the principal stress sum.

While the techniques described above, based on finite difference or analytic solutions are quite inventive, they generally lack the versatility of finite element based methods. Modern PC or workstation commercial FE systems provide a very high degree of flexibility. For example ANSYS contains a parametric modelling capability, coupled with a fairly complete macro language which, together with sophisticated graphical pre- and post-processing techniques, an optimisation module and the ability to include user-defined elements, makes it ideal (if not pre-eminent in the field) for hybrid experimental and analytical stress analysis. The imminent dilemma which must be faced is exactly how this capability is best used. The answer to this returns us to the initial question as to how the SPATE data should be used, and as argued previously, there are many answers to this.

The simplest, but emphatically non-trivial, use to which FE analysis could be used in conjunction with SPATE lies in the assessment of a FE model for design purposes. As discussed above, this use is not truly aimed at the stress separation problem, but this information could of course be derived from the analysis. It is fairly simple to carry out FE analysis, particularly with refined geometry modelling, automatic mesh generation and the coming adaptive meshing procedures. Arguably too easy! While modern FE packages are reasonably robust and quality assured, they are not infallible, and neither is the analyst. Yet, in many cases, an experienced analyst, involved in the stress analysis of a well-established family of industry-specific components, whose mechanical behaviour is well known can be reasonably confident in the results. But a radically new design, or a component whose behaviour is less well known would probably also need to be assessed by testing. The particular advantage of a full-field method like SPATE would be in assessing the adequacy of a FE model. This approach is hybrid in the sense of using experimental results to enhance a numerical analysis, rather than vice versa.

Examples of the use of an analysis system such as ANSYS depend on the imagination of the engineer, and even the simplest list would not be exhaustive. One extreme example which the writer has been developing is the use of SPATE coupled with the tomography module of the PDA Engineering PATRAN-II MCAE system. The tomography module of PATRAN, although designed for medical and engineering non-destructive evaluation (NDE) applications using CAT scans, can input any image scan. The conceptual modelling module of PATRAN can then be used to create geometry and guide mesh design in regions of high bulk-stress gradient.

### 9.3.3 The inverse problem

The inverse problem has been mentioned on several occasions in the preceding discussion without being fully defined. The simplest explanation of an inverse problem can be given in terms of a boundary value problem, such as the determination of displacement and stress in a deforming body of a specific material subject to applied boundary constraints and applied loads using the field equations of solid mechanics. With sufficient prescribed boundary conditions this problem is said to be a direct boundary value problem and well posed since it may be shown that a unique solution exists. This type of problem is familiar to most engineers.

The inverse problem involves the determination of unknown boundary conditions, say applied forces, given some measure of the solution of the problem in the interior, but not the complete solution. It is said to be ill-posed since a unique solution may not exist. Such problems can be found throughout mathematical physics (Romanov 1987). The usual practical approach to inverse ill-posed problems is the use of least-squares best-fit to the known interior data. However, it is well established that if the interior data derives from experimental data which is known to contain small errors, then the boundary solution may be unstable, that is, a small change in the interior data can produce a large change in the calculated boundary data. In this case some regularisation technique is usually required (the aim is smoothing, but the aims and methods used are a bit different).

A good discussion of the inverse problem in elasticity has been given recently (Maniatty *et al* 1989). This paper addresses the problem of determining unknown (or inexact) boundary conditions if measured displacements or strains are known in the interior, as may be obtained from interferometry or conventional strain-gauging. These authors examined a numerical simulation of a square plate with non-uniform applied stress estimated from measured interior displacements. It was shown that, if displacement data could be obtained close to the boundary, then the error in the calculated tractions was proportional to the error in the measured displacements. If the data could not be reliably obtained close to the boundary, then the error increased. These results are of course not directly applicable to the SPATE problem, since only scalar bulk stress is available, although the resolution is much greater than that considered by Maniatty and co-workers. However the technique used was quite general, based on FE analysis.

Considerable care is needed in using hybrid methods of experimental and numerical stress analysis in order to recognise an inverse problem. Some of the work cited in the preceding section on hybrid methods involves an inverse problem to some degree, depending on the experimental technique used.

The inverse problem has been considered for SPATE (Ryall and Wong 1988, Waldman *et al* 1990). Ryall and Wong carried out a numerical simulation on a square biaxial plate, using the measured bulk stress to estimate the elasticity solution through a least-squares fit with an unknown applied force. With suitable conditioning of the numerical solution, the results were excellent. This approach was extended by Waldman and co-workers to a simple three-dimensional problem (a cantilevered beam) using constrained FE techniques, again with good results. However no investigation of the stability of the solution, in terms of the derived boundary forces, was reported in either paper, although in the latter the so-called Fisher information matrix, which governs the variance of the estimated parameters in the least-squares fit, was examined. This can help with the *a priori* design of experiments, in this case helping to decide where SPATE data should be zoomed (or at least on spatial resolution), as similarly discussed by Maniatty and co-workers.

Clearly there is a lot of new work to be done if SPATE is to be used for the solution of inverse problems, and the studies by Ryall and co-workers represents a pioneering first step. The writer would expect a good deal of progress to be made with benchmark biaxial problems since, as mentioned above, SPATE does indeed supply interior data. As usual the three-dimensional problem is different, since SPATE does not supply interior data (that is, a partial solution to the stress analysis problem) but only boundary data. This type of problem has not been studied in the literature on inverse problems. The difficulty with inverse problems is that not only is a solution expected, but also a study of the stability of the solution, deriving from experimental error. It may be expected that this is geometry dependent, although again little work has been done for problems in elasticity. Here the writer takes the easy route, and simply points out the difficulties involved for future reference by the reader.

However, we are not done with inverse problems just yet. The type of inverse problem described above—the determination of uncertain boundary data—is not exclusive. Another type of inverse problem relates to parameter estimation, specifically the estimation of material properties from indirect experiments (Beck and Arnold 1977). Since the SPATE system makes specific assumptions about material properties for routine use, the reader may wonder of what use this observation is. But this type of inverse problem may be used to some effect: it may be used to assess the validity of the SPATE material assumptions from the test itself. For example, the value of Young's modulus (and Poisson's ratio) which best fits the SPATE data may be determined using raw data (and smoothed data). If the resulting estimates are not too different from each other and from the expected values of these parameters for the material, then methods of stress separation (or otherwise) can probably be used with an additional degree of confidence. If this is not the case, the regions of possible inhomogeneity could be identified and the procedure repeated with variations to the stiffness. An initial investigation of this type of procedure, using the ANSYS optimisation module which can form a design function based on a least-squares difference between calculated FE nodal results and SPATE bulk stress, has been made.

Other possibilities for inverse problems can be envisaged: for example the smoothing of the edge error in the bulk stress using the interior data. In the writer's view, the reader would do better to examine the limitations of the three-dimensional problem, before developing complex techniques for inverse problems in two dimensions.

## 9.4 THE EXTENSION TO THREE-DIMENSIONAL PROBLEMS

It is clear that a great deal of progress could be made with the biaxial problem using a variety of techniques addressing a variety of engineering problems. But, as stated in the introduction, the reader should always bear in mind the possible treatment of the 'real' three-dimensional problem.

As far as the writer is aware, the only successful stress separation which has been carried out for a three-dimensional problem is the torispherical pressure vessel head (Stanley 1989b). This used a special property of the stresses in thin shells of revolution to evaluate the surface stresses using a 'marching' technique with good success. The question is, what can be done for a general three-dimensional component, such as a casting? The answer demonstrates the limitations.

At the present state of development of the SPATE technique, little progress can be made with purely equilibrium methods. The reason is two-fold. Firstly, in three dimensions there are six components of the symmetric stress tensor and only three equations of equilibrium; while the bulk stress would only serve to remove one of these (and only on the boundary), some progress could possibly be made for special components using additional information on the stress components, as is the case for thin shells. Secondly, even if sufficient stress components could be eliminated for special cases, the equilibrium equations in curvilinear coordinates would contain derivatives of bulk stress normal to the surface. This information is of course unknown. The unconvinced should examine the counter example to uniqueness for a simple beam given in the paper by Waldman *et al* (1990).

Similarly the solution of three-dimensional inverse problems using SPATE is burdened with additional uncertainty since only surface data is available (although as demonstrated by Waldman *et al* (1990), some progress can be made in particular cases; the unfortunate aspect is that this may not be evident before the inverse least-squares analysis is carried out!).

This only leaves one viable approach, and fortunately it is the easiest and most logical.

## 9.5 DISCUSSION

The reader will have noticed that while the writer has much to say on the problem of post-processing SPATE data, there are few hard results. Indeed the reader can put himself in the role of the writer as a newcomer to this subject. In reviewing the work which has been done to date, many deceptive avenues have been examined, and later dropped as lacking generality; this reflects the confusion which the writer has perceived in the literature as a novice. There are many ways forward for biaxial problems; a complete summary of this will be given in the work of a doctoral student in the near future. There is only one way forward for the general three-dimensional problem: the use of SPATE in conjunction with a parallel finite element analysis which uses, and perhaps checks, the basic material assumptions and assumed boundary conditions, in an interactive way, refining the finite element model with re-analysis guiding subsequent SPATE scans.

In terms of progress in MCAE techniques, the way forward should include a high-performance engineering workstation. This would give the interactive computing power, coupled with enhanced display resolution and interactive real-time three-dimensional graphics (as on a Silicon Graphics Personal Iris), which would be necessary to make this a practicality. Fortunately the price performance ratio of such hardware can be expected to continue to drop dramatically in the near future. The question is whether it is thought worthy of the effort. In the writer's opinion it is: the capability is there, it should be used now.

It has been noted by many writers (Dally 1987a) that experimental mechanics has been losing ground to computational mechanics in the past decade, and that techniques such as photoelasticity have suffered the most. Dally considers that experimental mechanics should recover in the coming decade, probably with a shift of emphasis to a study of localised damage and failure mechanisms through micro-mechanics. The writer would agree that this is likely, particularly with the growing sophistication in finite element analysis, coupled with computational power, which can be expected to continue. For the majority of design situations the engineer will have increased confidence in the use of computational mechanics without resort to experimentation. SPATE has a clear role here in the rapid assessment of FE modelling, if this is deemed necessary. However, if the future of experimental mechanics does lie in failure assessment, then SPATE has a less obvious role unless there is more study of the relation between continuum thermodynamics and micro-mechanics, and the consequent interpretation of the basic SPATE signal.

## 9.6 ACKNOWLEDGMENTS

The writer would like to acknowledge continuing support from the National Engineering Laboratory through Extra Mural Research Agreements. Use of the ANSYS system was made available through an educational licence from Swanson Analysis Systems Inc, USA. Several of the methods proposed here are being developed as part of a doctoral thesis by Mr R Hamilton supported through a studentship by NEL.

# References

Arcan M and Bank-Sills L 1982 *Proc. 7th Int. Conf. on Experimental Stress Analysis (Haifa, August 1982)* pp 187–201

Ashton J E, Halpin J C and Petit P H 1969 *Primer on Composite Materials: Analysis* (Stamford: Technomic)

Balas J 1967 *Exp. Mech.* **7** 127–39

Balas J, Sladek J and Drzik M 1983 *Exp. Mech.* **23** 196–202.

Bank-Sills L, Arcan M and Gabey H 1984 *Eng. Fract. Mech.* **19** 739–50

Beck J V and Arnold K J 1977 *Parameter Estimation in Engineering and Science* (New York: Wiley)

Beck J V, Blackwell B and St Clair Jr C R 1985 *Inverse Heat Conduction: Ill-Posed Problems* (New York: Wiley)

Beghi M G and Bottani C E 1980 *Appl. Phys.* **23** 57–9

Beghi M G, Bottani C E and Caglioti G 1986 *Res. Mech.* **19** 365–79

Beghi M G, Bottani C E, Caglioti G and Fazzi A 1987a *Proc. 2nd Int. Conf. on Stress Analysis by Thermoelastic Techniques (London, February 1987)* ed B Gasper (Bellingham, WA: SPIE vol 731) pp 56–67

Beghi M G, Bottani C E and Fazzi A 1987b *J. Phys. E: Sci. Instrum.* **20** 26–32

Belgen M H 1967 *ISA Trans.* **6** 49–53

—— 1968 *NASA Cntr. Report* CR–1067

Bendat J S and Piersol A G 1986 *Random Data: Analysis and Measurement Procedures* (New York: Wiley) ch 1

Berghaus D G 1989 Combining photoelasticity and finite element methods for stress analysis using least squares. *Proc. SEM Spring Conf. on Experimental Mechanics (Boston, 1989)*

Bever M B, Holt H L and Titchener A L 1973 *The Stored Energy of Cold Work* (Oxford: Plenum); *Prog. Mater. Sci.* **17**

Biot M A 1956 *J. Appl. Phys.* **27** 240–53

Blotny R and Kaleta J 1986 *Int. J. Fatigue* **8** 29–33

Blotny R, Kaleta, J. Grzebien W and Adamczewski W 1986 *Int. J. Fatigue* **8** 35–8

Bottani C E and Caglioti G 1982a *Phys. Scr.* **T1** 65–70

—— 1982b *Mater. Lett.* **1** 119–21

Brammer J A and Percival C M 1970 *Exp. Mech.* **10** 245–50

Brandt S 1976 *Statistical and Computational Methods in Data Analysis* 2nd edn (Amsterdam: North-Holland)

Bream R G, Gasper B C, Lloyd B E and Everett G M 1989 *Proc. Int. Conf. on Stress and Vibration (London, March 1989)* ed P Stanley (Bellingham, WA: SPIE vol 1084) pp 279–92

Bridgman P W 1950 *Rev. Mod. Phys.* **22** 56–63

Brown K 1987 *Proc. 2nd Int. Conf. on Stress Analysis by Thermoelastic Techniques (London, February 1987)* ed B Gasper (Bellingham, WA: SPIE vol 731) pp 205–11

Burnay S G, Williams T L and Jones C H N (ed) 1988 *Applications of Thermal Imaging* (Bristol: Adam Hilger) ch 1

Carslaw H S and Jaeger J C 1959 *Conduction of Heat in Solids* (Oxford: Clarendon)

Cavaco M A M and de Freire J L 1989 A computer based analytic and photoelastic method for stress analysis *Proc. 10th Brazilian Congress on Mechanical Engineering (Rio de Janeiro, 1989)*

Cernocky E P and Krempl E 1981 *J. de Mecan. Appliquée* **5** 293–321

Chalmers G F 1980 *Proc. BSSM Conf. on Product Liability and Reliability (Birmingham, UK, 1980)* (Newcastle: BSSM) paper 5A

Chambless D *et al* 1986 A new hybrid photoelastic finite element technique for stress analysis *Proc. Spring SEM Conf. on Experimental Mechanics (New Orleans, 1986)*

Chan S W K and Tubby P J 1988 *Welding Inst. Report* 369/1988 pp 1–24

Chisholm D B and Jones D J 1977 *Expl. Mech.* **17** 7–13

Christensen R M 1971 *Theory of Viscoelasticity: An Introduction* (New York: Academic)

—— 1979 *Mechanics of Composite Materials* (New York: Wiley)

Chrysochoos A 1985 *J. Mecan. Theor. Appliquée* **4** 589–614

Compton K T and Webster D B 1915 *Phys. Rev.* **5** 159–66

Cottrell A H 1953 *Dislocations and Plastic Flow* (Oxford: Clarendon)

Cummings W M and Harwood N 1985 *Proc. Int. Conf. on Experimental Mechanics (Las Vegas, June 1985)* (Bethel, CT: SEM) pp 740–46

—— 1987 *Proc. Conf. on Optical and Opto-electronic Applied Science and Engineering (San Diego, August 1987)* (Bellingham, WA: SPIE vol 817) pp 96–108

Dally J W 1987a *Review of current developments in experimental mechanics. Applied Solid Mechanics – 2* ed J Spence and A S Tooth (Applied Science)

—— 1987b *Photoelasticity. Handbook on Experimental Mechanics* ed AS Kobayashi (Englewood Cliffs, NJ: Prentice Hall)

Darken L and Curry R 1953 *Physical Chemistry of Metals* (London: McGraw-Hill) ch 7

de Groot S R and Mazur P 1962 *Non-equilibrium Thermodynamics* (Amsterdam: North-Holland)

Dennis P N J 1986 *Photodetectors: An Introduction to Current Technology* (New York: Plenum) ch 1

Dillon O W and Taucher T R 1966 *Int. J. Solids Struct.* **2** 385–91

Dixon M 1988 *Strain* **24** 139–42

Drucker D C 1967 *Introduction to Mechanics of Deformable Solids* (New York: McGraw-Hill) p 71

Duncan J L 1988 A method of stress analysis on fresh bone by thermoelastic emission *PhD Thesis* University of Strathclyde, Glasgow

Dunn S A, Lombardo D and Sparrow J S 1989 *Proc. Int. Conf. on Stress and Vibration (London, March 1989)* ed P Stanley (Bellingham, WA: SPIE vol 1084) pp 129–42

Enke N F (ed) 1988 *Theory and Applications of Thermographic Stress Analysis* (Madison, WI: University of Wisconsin)

—— 1989 *Proc. Int. Conf. on Stress and Vibration (London, March 1989)* ed P Stanley (Bellingham, WA: SPIE vol 1084) pp 84–102

Ewins D J 1984 *Modal Testing: Theory and Practice* (Letchworth, UK: Research Studies Press) pp 174–80

Farlow S J 1982 *Partial Differential Equations for Scientists and Engineers* (New York: Wiley)

Farren W S and Taylor G I 1925 *Proc. R. Soc.* A **107** 422–51

Feng Z and Rowlands R E 1987 *Comput. Struct.* **26** 979–90

Fisher M J and Herrmann G 1984 *Rev. Prog. Quant. NDE* B **3** 1283–91

Friedel J 1964 *Dislocations* (Oxford: Pergamon)

Fung Y C 1965 *Foundations of Solid Mechanics* (Englewood Cliffs, NJ: Prentice Hall)

Gilbert J A *et al* 1987 Two dimensional stress analysis combining high frequency Moire measurements and finite element modelling *Experimental Techniques* (Bethel, CT: SEM)

Gipson G S 1989 Basic concepts and recent developments in the Poisson equation *Computational Mechanics* (Berlin: Springer)

Gough J 1805 *Manchester Phil. Mem.* **2** 288–95

Haga H 1982 *Ann. Phys. Chem.* **15** 1–18

Hammersley J M and Handscomb D C 1964 *Monte Carlo Methods* (London: Methuen)

Harwood N 1988 *Strain* **24** 67–70

Harwood N and Cummings W M 1987 *Proc. 5th Int. Modal. Analysis Conf. (London, April 1987)* vol 1 (Schenectady, NY: Union College) pp 399–405

—— 1989 *Proc. Int. Conf. on Stress and Vibration (London, March 1989)* ed P Stanley (Bellingham, WA: SPIE vol 1084) pp 143–58

Hayward A T J 1977 *Repeatability and Accuracy* (London: Mech. Eng. Publ.)

Hirao M and Pao Y H 1985 *J. Acoust. Soc. Am.* **77** 1659–64

Hirth J P and Lothe J 1968 *Theory of Dislocations* (New York: McGraw-Hill)

Huang Y M, Hamdi-Abdelmohsen H H, Lohr D, Rowlands R E and Stanley P 1988 *Proc. 6th Int. Cong. on Experimental Mechanics (Portland, OR, June 1988)* (Bethel, CT: SEM) pp 578–84

Hudson R D 1969 *Infra-red System Engineering* (New York: Wiley) p 294

Jones R, Tay T E and Williams J F 1987 Thermomechanical behaviour of composites *Proc. US Army Workshop on Composite Material Response: Constitutive Relations and Damage Mechanics (Glasgow, July 1987)* ed G C Sih, G F Smith, I H Marshall and J J Wuh (New York: Elsevier) pp 49–59

Jordan E H and Sandor B I 1978 *J. Test. Eval.* **6** 325–31

Joule J P 1859 *Phil. Trans.* **149** 91–130

Kassir M K and Sih G C 1975 *Three-dimensional Crack Problems – Mechanics of Fracture* vol 2 (Leyden: Nordhoff)

Kobayashi A S 1987 Hybrid experimental stress analysis *Handbook on Experimental Mechanics* ed A S Kobayashi (Englewood Cliffs, NJ: Prentice-Hall)

Krempl E 1985 *Plasticity Today* ed A Sawczuk and G Bianchi (London: Elsevier) ch 14

Lancaster P 1986 *Curve and Surface Fitting: An Introduction* (New York: Academic)

Landau L D and Lifshitz E M 1970 *Theory of Elasticity* 2nd edn (Oxford: Pergamon) ch 1

Lesniak J R 1988 *Proc. 6th Int. Cong. on Experimental Mechanics (Portland, OR, June 1988)* (Bethel, CT: SEM) pp 825–9

Loader A J, Turner W B and Harwood N 1987 *Proc. 2nd Int. Conf. on Stress Analysis by Thermoelastic Techniques (London, February 1987)* ed B Gasper (Bellingham, WA: SPIE vol. 731) pp 149–53

Machin A S, Sparrow J G and Stimson M G 1987 *Strain* **23** 27–30

Maniatty A, Zabaras N and Stelson K 1989 *J. Eng. Mech.* **115** 1303–17

MacKenzie A K 1989 *Proc. Int. Conf. on Stress and Vibration (London, March 1989)* ed P Stanley (Bellingham, WA: SPIE vol 1084) pp 59–71

McKelvie J 1987 *Proc. 2nd Int. Conf. on Stress Analysis by Thermoelastic Techniques (London, February 1987)* ed B Gasper (Bellingham, WA: SPIE vol 731) pp 44–53

*Metals Handbook* 1980 9th edn (Metals Park, Ohio: American Society of Metals)

Mitchell L D, Deel J C, Cobb R E and Luk Y W 1987 *Proc. 5th Int. Modal Analysis Conf. (London, April 1987)* vol 1 (Schenectady, NY: Union College) pp 364–73

Moslehy F A and Ranson W F 1980 Laser speckle interferometry and boundary integral techniques in experimental stress analysis *Developments in Theoretical and Applied Mechanics* ed J E Stoneking (Tennessee) pp 473–92

Mountain D S and Cooper G P 1989 *Strain* **25** 15–19

Mountain D S and Webber J M B 1978 *Proc. Soc. Photo-Opt. Instrum. Eng.* **164** 189–96

Nabarro F R N 1967 *Theory of Crystal Dislocations* (Oxford: Clarendon)

Nayroles B, Bouc R, Caumon H and Chezaux J C 1981 *Int. J. Eng. Sci.* **19** 929–47

Nowick A S and Berry B S 1972 *Anelastic Relaxation in Crystalline Solids* (New York: Academic)

Numerical Algorithms Group: Library Introductory Guide. NAG, 1987

Oliver D E, Razdan D and White M T 1982 Structural design assessment using thermoelastic stress analysis (TSA) *Proc. BSSM/RAeS Conf.* (Newcastle: BSSM)

Oliver D E and Webber J M B 1984 *Proc. 5th Int. Cong. on Experimental Mechanics (Montreal, June 1984)* (Washington, DC: SESA) pp 539–46

Otnes R K and Enochson L 1978 *Applied Time Series Analysis* (New York: Wiley) p 290

Paris P C and Erdogan F 1963 *Trans. ASME J. Basic Eng.* **85** 528–34

Potter R T 1987 *Proc. 2nd Int. Conf. on Stress Analysis by Thermoelastic Techniques (London, February 1987)* ed B Gasper (Bellingham, WA: SPIE vol 731) pp 110–20

Potter R T and Greaves L J 1987 *Proc. Conf. on Optical and Opto-electronic Applied Science and Engineering (San Diego, August 1987)* (Bellingham, WA: SPIE vol. 817) pp 134–46

Pukas S R 1987 *Proc. 2nd Int. Conf. on Stress Analysis by Thermoelastic Techniques (London, February 1987)* ed B Gasper (Bellingham, WA: SPIE vol 731) pp 88–101

Randall R B 1987 *Frequency Analysis* 3rd edn (Naerum, Denmark: Bruel and Kjaer) p 233

Richard H A 1980 *Int. J. Fracture* **17** 105–7

Riley K F 1974 *Mathematical Methods for the Physical Sciences* (Cambridge: Cambridge University Press)

Rocca R and Bever M B 1950 *Trans. AIME* **188** 327–33

Rocklin T G, Crowley J and Vold H 1984 *Proc. 3rd Int. Modal Analysis Conf. (Orlando, 1984)* (Schenectady, NY: Union College) pp 272–8

Rogers G F C and Mayhew Y R 1980 *Engineering Thermodynamics, Work and Heat Transfer* 3rd edn. (London: Longman) ch 23

Romanov V G 1987 *Inverse Problems of Mathematical physics* (VNU Science Press)

Ronnpagel D 1979 *J Phys. E: Sci. Instrum.* **12** 409–17

Ronnpagel D and Schwink C 1978 *Acta Metall.* **26** 319–31

Rooke D P and Cartwright D J 1976 *Compendium of Stress Intensity Factors* (London: HMSO)

Rowlands R E 1986 Separating stresses from SPATE data *SPATE User's Workshop, SEM Meeting (New Orleans, 1986)*

Ryall T G, Cox P M and Enke N F 1990 *Mech. Systems Signal Processing* to be published

Ryall T G and Wong A K 1988 *Mech. Mater.* **7** 205–14

Saul R S and Webber J M B 1987 *Proc. Thermosense IX (Orlando, May 1987)* ed R P Madding (Bellingham, WA: SPIE vol 780)

Smith R A 1979 *Fracture Mechanics: Current Status, Future Prospects* (Cambridge: Pergamon)

Stanley P 1986 Stress separation *SPATE User's Workshop, SEM Spring Conf. (New Orleans, 1986)*

—— 1989a *Proc. Int. Conf. on Stress and Vibration (London, March 1989)* ed P Stanley (Bellingham, WA: SPIE vol 1084) pp 72–83

—— 1989b Stress separation from SPATE data for a rotationally symmetrical pressure vessel *Applied Solid Mechanics – 3* ed I M Allison and C Ruiz (Amsterdam: Elsevier)

Stanley P and Chan W K 1985a *Proc. Int. Conf. on Experimental Mechanics (Las Vegas, June 1985)* (Bethel, CT: SEM) pp 747–757

—— 1985b *J. Strain Analysis* **20** 129–37

—— 1986a *Proc. Int. Conf. on Fatigue of Engineering Materials and Structures (Sheffield, September 1986)* vol 1 (London: Institute of Mechanical Engineers) pp 105–14

—— 1986b *Proc. Int. Conf. on Experimental Mechanics (New Orleans, June 1986)* (Bethel, CT: SEM) pp 916–23

—— 1987a *Proc. 2nd Int. Conf. on Stress Analysis by Thermoelastic Techniques (London, February 1987)* ed B Gasper (Bellingham, WA: SPIE vol 731) pp 102–9

—— 1987b *Proc. 2nd Int. Conf. on Stress Analysis by Thermoelastic Techniques (London, February 1987)* ed B Gasper (Bellingham, WA: SPIE vol 731) pp 17–25

Swanson Analysis Systems: ANSYS PC/THERMAL User's Guide 1987

Tamman G and Warrentrup H 1937 *Ztsch. Metal.* **29** 84–8

Taylor G I and Quinney H 1934 *Proc. R. Soc.* A **143** 307–26

Tennant R M 1971 *Science Data Book* (Edinburgh: Oliver and Boyd) p 60

Thompson J C, Abou el Atta H 1975 On the application of complex potentials and photoelastic data to the solution of plane problems *J. BSSM* **11**

Thomson W 1853 *Trans. R. Soc. Edinburgh* **20** 261–83

—— 1878 *Encyclopaedia Britannica* 9th edn 813–16

Thomson W T 1972 *Theory of Vibration* (Englewood Cliffs, NJ: Prentice-Hall) p 23

Timoshenko S 1956 *Strength of Materials* (Princeton, NJ: Van Nostrand)

Timoshenko S P and Goodier J N 1982 *Theory of Elasticity* (Tokyo: McGraw-Hill) p 456

Turner S R and Pollard N G 1987 *Proc. 2nd Int. Conf. on Stress Analysis by Thermoelastic Techniques (London, February 1987)* ed B Gasper (Bellingham, WA: SPIE vol 731) pp 162–77

Waldman W, Ryall T G and Jones R 1990 On the determination of stress components in 3D from thermoelastic data. To appear

Wallace D C 1972 *Thermodynamics of Crystals* (New York: Wiley) ch 1

Weathers J M *et al* 1985 *Exp. Mech.* **25** 60–5

Webber J M B 1987 *Proc. 2nd Int. Conf. on Stress Analysis by Thermoelastic Techniques (London, February 1987)* ed B Gasper (Bellingham, WA: SPIE vol 731) pp 2–16

—— 1988 Private communication

Weber W 1830 *Ann. Phys. Chem.* **20** 177–213

Williams R O 1960 *Rev. Sci. Instrum.* **31** 1336–41

——1961 *Acta Metall.* **9** 949–57

——1963 *Rev. Sci. Instrum.* **34** 639–43

——1964 *Acta Metall.* **12** 745–57

——1967 *Advances in Materials Research* ed. H Herman vol 1 (New York: Interscience) pp 251–78

Wolfenden A 1968 *Acta Metall.* **16** 975–80

Wolfenden A and Appleton A S 1967 *Rev. Sci. Instrum.* **38** 826–30

Wong A K 1990 A non-adiabatic thermoelastic theory for composite laminates *Royal Aircraft Establishment Technical Report* TR90007 (Also to be published in *J. Phys. Chem. Solids*)

Wong A K, Dunn S A and Sparrow J G 1988b *Nature* **332** 613–15

Wong A K, Jones R and Sparrow J G 1987 *J. Phys. Chem. Solids* **48** 749–53

Wong A K, Sparrow J G and Dunn S A 1988a *J. Phys. Chem. Solids* **49** 395–400

Wood A, Hurden D C and Forsyth P 1987 *Proc. 2nd Int. Conf. on Stress Analysis by Thermoelastic Techniques (London, February 1987)* ed B Gasper (Bellingham, WA: SPIE vol 731) pp 154–61

Woolman J and Mottram A 1964 *The Mechanical and Physical Properties of the British Standard En Steels* vol 1 (Oxford: Pergamon)

Zener C 1938 *Phys. Rev.* **53** 90–9

Zhang D and Sandor B I 1989 *Proc. 17th Symp. on NDE (April 1989)* ed. F A Iddings and D W Moore (San Antonio: Am. Soc. for NDT) pp 155–66

# Index